Transformation Process and Toxic Effects of Pollutants in Agricultural Environment

Transformation Process and Toxic Effects of Pollutants in Agricultural Environment

Guest Editors

Changbo Zhang
Liang Peng
Weijie Xue

MDPI

Basel • Beijing • Wuhan • Barcelona • Belgrade • Novi Sad • Cluj • Manchester

Guest Editors

Changbo Zhang
Agro-Environmental Protection Institute
Ministry of Agriculture and Rural Affairs
Tianjin
China

Liang Peng
College of Environment and Ecology
Hunan Agricultural University
Changsha
China

Weijie Xue
Agro-Environmental Protection Institute
Ministry of Agriculture and Rural Affairs
Tianjin
China

Editorial Office
MDPI AG
Grosspeteranlage 5
4052 Basel, Switzerland

This is a reprint of the Special Issue, published open access by the journal *Toxics* (ISSN 2305-6304), freely accessible at: https://www.mdpi.com/journal/toxics/special_issues/042Q1611H0.

For citation purposes, cite each article independently as indicated on the article page online and as indicated below:

Lastname, A.A.; Lastname, B.B. Article Title. *Journal Name* **Year**, *Volume Number*, Page Range.

ISBN 978-3-7258-7284-8 (Hbk)
ISBN 978-3-7258-7285-5 (PDF)
https://doi.org/10.3390/books978-3-7258-7285-5

Contents

About the Editors

Changbo Zhang

Changbo Zhang is an associate researcher at the Agro-Environmental Protection Institute of the Ministry of Agriculture and Rural Affairs of China. He received a Doctor of Science degree from Wuhan University in 2009 and a Doctor of Science degree in Chemistry from the University of Clermont-Ferrand II in France in 2010. He is a master's supervisor at the Chinese Academy of Agricultural Sciences. He is a member of the 7th Council of the Tianjin Environmental Science Society, a member of the Tianjin Science and Technology Expert Database, a soil pollution prevention expert of the Tianjin Environmental Protection Bureau, an agricultural finance project expert of the Hebei Provincial Department of Agriculture and Rural Affairs, and a high-level expert of the Science and Technology Innovation Think Tank of the Shandong Science and Technology Consulting Association. He has been engaged in research on the transfer mechanisms of pollutants in the environment and agricultural environmental pollution and remediation. He has systematically studied the ecological toxicological chemistry of heavy metals in rice and wheat, as well as the process mechanisms of advanced oxidation treatment technologies for emerging contaminants and their impact mechanisms on the chemical transformation process of heavy metals in the environment. He has led 12 projects, including the National Scholarship Foundation of China, National Natural Science Foundation of China, Natural Science Foundation of Tianjin, Key National Research and Development Program Projects, the Foreign Experts Project of the Ministry of Science and Technology of China, and the Foreign Experts Project of the Ministry of Human Resources and Social Security of China. He has registered three products of heavy metal leaf surface control agents and has applied for and obtained more than 50 patents. He has published over 120 research papers.

Liang Peng

Liang Peng is a professor and doctoral supervisor at Hunan Agricultural University; member of the National Professional Committee for Heavy Metal Pollution Prevention and Control. His research area includes the functions and applications of soil crusts in heavy metal-contaminated farmland and the environmental remediation technology. He has undertaken more than 20 national-level key research and development projects and projects supported by the National Natural Science Foundation of China. He has published over 60 SCI papers in internationally renowned journals such as the *Chemical Engineering Journal* and *Journal of Hazardous Materials*. He has obtained nine invention patent authorizations, transferred three invention patents, and realized the transformation of two cadmium-reducing foliar fertilizer products. Additionally, he has been awarded the Third Prize of the Hunan Provincial Natural Science Award, as well as the First and Second Prizes of the Changsha Municipal Academic Paper Award.

Weijie Xue

Weijie Xue is an associate researcher at the Agro-Environmental Protection Institute, Ministry of Agriculture and Rural Affairs of China. She holds a Ph.D. in Plant Nutrition from the Chinese Academy of Agricultural Sciences. Her main research directions include environmental pollution and remediation, pollution control and environmental remediation technologies, and molecular mechanisms of pollutant migration. She has led nine projects, including those funded by the National Natural Science Foundation of China, the State Administration of Foreign Experts Affairs (Personal Program), and provincial/ministerial-level initiatives. She is the first cohort of Tianjin

Young Scientific and Technological Talents, 2024. In recent years, she has been engaged in research on the ecotoxicology of heavy metals in plants and the mechanisms of heavy metal response in plant cells. Focusing on the issue of cadmium (Cd) contamination in rice, she has conducted key research on the molecular mechanisms of Cd uptake and transport in rice. Research results, as first author or corresponding author, include over 20 SCI-index papers published in JCR Q1 journals such as the *Journal of Hazardous Materials*, *Ecotoxicology and Environmental Safety*, and *Environmental Technology Innovation*. One of these papers was awarded a provincial-level First Prize for Outstanding Scientific Paper. The applicant holds eight authorized patents. Two Cd-reducing foliar fertilizers developed have been officially registered. Previous achievements contributed to the applicant receiving, as a key team member, the Chinese Academy of Agricultural Sciences Achievement Transformation Award - Science and Technology Development Award (Team Award) and the DBN Collaborative Innovation Award (Team Award). The findings on GLR-mediated heavy metal uptake and transport in rice served as the primary supporting material for the appraisal of one internationally leading scientific achievement.

toxics

Editorial

Transformation and Toxic Effects of Pollutants in Agricultural Environment

Changbo Zhang [1], Liang Peng [2,*] and Weijie Xue [1,*]

1 Agro-Environmental Protection Institute, Ministration of Agriculture and Rural Affairs, Tianjin 300191, China; zhangchangbo@caas.cn
2 College of Environment and Ecology, Hunan Agricultural University, Changsha 410128, China
* Correspondence: pengliang2004@126.com (L.P.); tzfelicity@163.com (W.X.)

1. Introduction

The agricultural environment is of critical for human survival on Earth, especially as the water–soil–crop agricultural environment system is closely related to human health [1]. Pollutants such as heavy metals [2,3], microplastics, and pesticides [4–6] enter the agricultural environment through agricultural production, posing threats to agricultural safety and seriously affecting human health. Therefore, it is critical to explore the transformation and migration of pollutants in the agricultural environment and their toxicity changes and conduct research on related pollution remediation technologies to protect the quality of the agricultural environment.

2. An Overview of the Published Articles

Soil heavy-metal pollution poses a significant threat to soil quality and human health. Radina Nikolova et al. (contribution 1) evaluated the diversity and structural changes of microbial communities and assessed the ecological risks posed by different soils contaminated with heavy metals while predicting the key metabolic pathways related to soil microbial resistance. They emphasized the necessity of including soil microbial indicators in agricultural management strategies to ensure food safety. The impact of microplastics on the bioavailability of heavy metals is a current research hotspot. Zhenbo Wang et al. (contribution 2) conducted soil culture experiments to investigate the effects of polyethylene microplastic concentrations on soil properties, bacterial communities, surface chemical properties, and the distribution of forms of cadmium and lead in the soil, revealing that the concentration of polyethylene particles changed the physical and chemical properties of the soil and the structure of the bacterial community, ultimately affecting the transformation of forms of cadmium and lead.

Seeking to reduce the toxicity of heavy metals, Fu Lin et al. (contribution 3) conducted field experiments on double-cropping rice, explaining that spraying ionic liquids on the surfaces of the leaves of rice plants can inhibit the absorption, transport, accumulation, and toxicity of cadmium and arsenic in rice grains by promoting the synthesis of amino acids and regulating the absorption and transportation of essential elements. The results of field experiments conducted by Chen Kexin et al. (contribution 4) showed that spraying citric acid on the surfaces of rice plant leaves significantly reduced the soluble cadmium content in the plant's organs and promoted the transformation of cadmium from soluble into non-soluble forms, thereby inhibiting the transport of cadmium from plant organs to rice grains. Ge Lei et al. (contribution 5) evaluated the key sites for controlling cadmium transfer in rice through leaf surface control technology and revealed the good blocking

 https://doi.org/10.3390/toxics14030252

function of the endophytic microbial community biofilm with respect to the transport of heavy metals in rice. Zhang Shuo et al. (contribution 6) explored the potential of malic acid in reducing cadmium toxicity and investigated its genotype-dependent influence on cadmium absorption and essential element balance in rice. These findings provide a mechanistic basis for developing leaf-surface-conditioning agents based on micro-addition and formulating management strategies for heavy-metal pollution targeting different genotypes. These results provide technical support and a theoretical basis for reducing the excessive cadmium content in rice.

Soil pesticide residues constitute another widely concerning pollution problem. Jiale Zhang et al. (contribution 7) developed a method for microbially degrading new neonicotinoid insecticides in soil, studied the ability of Aspergillus versicolor in soil to degrade acetamiprid, and determined the optimal degradation conditions for and conducted toxicological analysis of acetamiprid and its metabolites. These findings will aid researchers in conducting safety assessments of the toxicological properties of neonicotinoid insecticides and their biodegradable metabolites, along with related studies on their degradation capabilities. Luan Gabriel Xavier de Souza et al. (contribution 8) investigated the adsorption of atrazine on pristine and aged polyethylene microplastics and evaluated its impact on plant toxicity. The results showed that the coexistence of microplastics and atrazine led to greater toxicity, indicating that the synergistic effect of pollutants can amplify the negative impact on plant development. Yuan Zhang et al. (contribution 9) reported that machine learning techniques are becoming increasingly valuable in simulating the transport of pollutants in plant systems and applied three symbolic regression models, based on original and enhanced data, to obtain accurate predictive equations. The results showed that machine learning models and symbolic regression methods can provide crucial insights into pollutants' absorption and accumulation in plant roots.

3. Conclusions

This Special Issue focuses on the transfer process and influence mechanisms of agricultural environment pollutants in the water–soil–crop system and explores the toxicity changes in heavy metals, microplastics, pesticides, and perfluoroalkyl and polyfluoroalkyl substances in the soil during their transfer in agricultural systems. It reports on leaf surface control technology and the way it is used to reduce excessive heavy metal concentrations in rice, along with the screening effect of heavy metals on soil microbe species. It introduces a machine learning model designed to reveal the synergistic effects of microplastics to enhance their toxicity and explains the mechanism by which microplastics affect the bioavailability of cadmium and lead in the soil. In summary, the articles published in this Special Issue cover the behavior and processes of pollutants in the agricultural environment; the diffusion mechanisms of pollutants; the factors affecting the distribution, migration, and transformation processes in the soil–crop system; and health risk assessments in the agricultural environment.

Data Availability Statement: No new data were created or analyzed in this study. Data sharing is not applicable to this article.

Acknowledgments: The Guest Editor is grateful to the editorial team, authors, and reviewers who contributed to the publication of this Special Issue. First, we would like to thank all the authors who submitted their work. We appreciate your trust, support, and contributions. Additionally, we sincerely thank the editorial team for their hard work and continuous administrative and technical support during the publication of the Special Issue. Finally, we would like to thank all the experts and scholars who participated in the review of the articles. We are grateful for your strict review of the submitted manuscripts, a process that has significantly improved the quality of the manuscripts.

https://doi.org/10.3390/toxics14030252

Conflicts of Interest: The authors declare no conflicts of interest.

List of Contributions:

1. Nikolova, R.; Gatev, E.; Kenarova, A.; Petkova, M.; Dinev, N.; Baldrian, P.; Radeva, G. Selective pressure of heavy metals on soil microbial taxa near a smelting area. *Toxics* **2025**, *13*, 1025.
2. Wang, Z.; Liu, S.; Zhao, P.; Li, G.; Duan, R.; Li, C.; Fu, H. Concentration-dependent effects of polyethylene microplastics on cadmium and lead bioavailability in soil. *Toxics* **2025**, *13*, 901.
3. Fu, L.; Deng, J.; Lao, D.R.; Zhang, C.; Xue, W.; Deng, Y.; Luo, X. Effects of foliar spraying of dicarboxylicdimethylammonium chloride on cadmium and arsenic accumulation in rice grains. *Toxics* **2024**, *12*, 418.
4. Chen, K.; Yu, B.; Xue, W.; Sun, Y.; Zhang, C.; Gao, X.; Zhou, X.; Deng, Y.; Yang, J.; Zhang, B. Citric acid inhibits Cd absorption and transportation by improving the antagonism of essential elements in rice organs. *Toxics* **2024**, *12*, 431.
5. Lei, G.; Song, H.; Gan, Z.; Yang, Y.; Chen, A. Foliar transpiration inhibitor reduces Cd accumulation in rice grain: The potential effect of the endophytic bacterial community. *Toxics* **2025**, *13*, 755.
6. Zhang, S.; Zhang, Y.; Lv, G.; Liu, T.; Liu, Z.; Jiang, Y.; Hao, Y.; Yu, Y.; Dong, W.; Qian, C. Effects of malic acid on cadmium uptake and translocation and essential element accumulation in rice. *Toxics* **2025**, *13*, 811.
7. Zhang, J.; Wang, X.; Yue, W.; Bao, J.; Yao, M.; Ge, L. Toxicological analysis of acetamiprid degradation by the dominant strain Md2 and its effect on the soil microbial community. *Toxics* **2024**, *12*, 572.
8. Gabriel Xavier de Souza, L.; Javier Cuba Teran, F.; Medici Frayne Cuba, R.; Rodrigues Chaves, A.; Cristina da Silva, K. Interaction of microplastics with emerging organic pollutants: A study on atrazine adsorption and phytotoxicity. *Toxics* **2025**, *13*, 257.
9. Zhang, Y.; Li, Y.; Li, Y.; Zhao, L.; Yang, Y. Interpretable machine learning models and rymbolic regressions reveal transfer of per- and polyfluoroalkyl substances (PFASs) in plants: A new small-data machine learning method to augment data and obtain predictive equations. *Toxics* **2025**, *13*, 579.

References

1. Dai, W. Research on Prevention and Control of Chinese Agricultural Ecological Environment Pollution to Ensure Food Safety. *Adv. Mater. Res.* **2012**, *616–618*, 2247–2250. [CrossRef]
2. Fu, L.; Deng, J.; Zhang, C.; Xue, W.; Deng, Y.; Luo, X.; Liu, S.; Chen, K.; La O, D.R.; Mailhot, G.; et al. Simultaneous inhibition of Cd and As absorption and transport in rice via coordinated cell wall sequestration, transporter regulation, and chelating ligand synthesis. *Ecotoxicol. Environ. Saf.* **2026**, *309*, 119528. [CrossRef] [PubMed]
3. Lin, J.; Tong, F.; Chen, W.; Hu, X.; Zhou, D.; Gu, X. Simultaneous immobilization of arsenic and cadmium in paddy soil with co-application of birnessite and lime. *J. Hazard. Mater.* **2026**, *502*, 141015. [CrossRef] [PubMed]
4. Jin, T.; Tang, J.; Lyu, H.; Wang, L.; Gillmore, A.; Schaeffer, S. Activities of microplastics (MPs) in agricultural soil: A review of MPs pollution from the perspective of agricultural ecosystems. *J. Agric. Food Chem.* **2022**, *70*, 4182–4201. [CrossRef] [PubMed]
5. Huang, Y.; Liu, Q.; Jia, W.; Yan, C.; Wang, J. Agricultural plastic mulching as a source of microplastics in the terrestrial environment. *Environ. Pollut.* **2020**, *260*, 114096. [CrossRef] [PubMed]
6. Wu, C.; Pan, S.; Shan, Y.; Ma, Y.; Wang, D.; Song, X.; Hu, H.; Ren, X.; Ma, X.; Cui, J.; et al. Microplastics mulch film affects the environmental behavior of adsorption and degradation of pesticide residues in soil. *Environ. Res.* **2022**, *214*, 114133. [CrossRef] [PubMed]

https://doi.org/10.3390/toxics14030252

toxics

MDPI

Article

Selective Pressure of Heavy Metals on Soil Microbial Taxa near a Smelting Area

Radina Nikolova [1], Evan Gatev [1], Anelia Kenarova [2], Michaella Petkova [1], Nikolai Dinev [3], Petr Baldrian [4,*] and Galina Radeva [1,*]

1 Roumen Tsanev Institute of Molecular Biology, Bulgarian Academy of Sciences, Acad. G. Bonchev Str., Bl. 21, 1113 Sofia, Bulgaria
2 Department of Ecology and Environmental Protection, Faculty of Biology, Sofia University "St. Kl. Ohridski", 8 Dragan Tsankov Blvd., 1164 Sofia, Bulgaria
3 N. Poushkarov Institute of Soil Science, Agrotechnologies and Plant Protection, Agricultural Academy, 7 Shosse Bankya St., 1331 Sofia, Bulgaria
4 Laboratory of Environmental Microbiology, Institute of Microbiology of the Czech Academy of Sciences, Videnská 1083, 142 00 Prague 4, Czech Republic
* Correspondence: baldrian@biomed.cas.cz (P.B.); gradeva@bio21.bas.bg (G.R.); Tel.: +359-887686237 (G.R.)

Abstract

Soil pollution by heavy metals (HMs) poses a major threat to soil quality and human health, with mining and smelting industries identified as key sources. Soils around smelters are often considered polluted hotspots, being generally unsuitable for agricultural activities. Although many studies have identified microbial taxa able to survive in such environments, most have focused on relatively low HM concentrations. The purpose of the study was to assess the ecological risk and to evaluate the diversity and structural shifts in microbial communities, as well as to predict key metabolic pathways associated with HM resistance in soils near Pb–Zn smelter in Bulgaria. The soils ranged from low-risk to disastrous, with cadmium (Cd) identified as the primary contributor to soil toxicity. High-throughput sequencing of 16S rRNA and ITS amplicons revealed widespread dominance of the phyla Proteobacteria, Actinobacteriota and Acidobacteriota, and Ascomycota, with the prevailing classes Acidobacteriae, Chloroflexia, and Eurotiomycetes, indicating their high tolerance to HMs. Functional predictions suggested enrichment of key pathways in the most polluted soils related to HM resistance, including efflux systems and detoxifying enzymes. These results highlight the necessity of integrating soil microbial indicators into agricultural management strategies to ensure safe food production.

Keywords: heavy metals; soil pollution; bacterial diversity; fungal diversity; bacterial resistance; metagenomics; functional profiling

1. Introduction

Soil pollution with HMs originating from anthropogenic activities, such as mining, smelting, foundries, and other industrial processes, poses a significant threat to soil quality and fertility, plant productivity, and human health [1]. Chronic exposure to elevated concentrations of HMs affects soil microorganisms, plants, and animals, both through direct contact with contaminated soils and via the trophic transfer of pollutants through food chains. Microbial communities, as vital components of soil, are essential to many ecological processes, including organic matter degradation and regulation of biogeochemical cycles, soil structure formation and stabilization, as well as suppression of plant pathogens [2].

Moreover, microbial communities respond immediately to HM stress, and their structural and functional responses make them reliable bioindicators of soil pollution and ecological disturbance [3]. Numerous studies have shown that HMs significantly reduce microbial diversity [4,5], richness [6], biomass and metabolic activity [5,6], and community structure and function [7–9].

The use of high-throughput sequencing technologies has significantly enhanced our understanding of microbial communities across various environments [10,11]. Metagenomic studies show that long-term HM exposure promotes tolerant taxa such as Proteobacteria, Actinobacteriota [6,12,13], Firmicutes [14], Gemmatimonadota and Chloroflexota [15,16]. It is also reported that the structure of soil fungal communities changes significantly depending on the level and type of HM pollution [17,18], and the dominant phyla in most of the studied soils are Ascomycota [19,20] and Zygomycota [21].

In general, microbial communities exposed to chronic HM pollution adapt by enhancing the activity of dominant and key taxa, reflecting their metabolic plasticity and functional redundancy [22,23]. Recent studies have shown that under HM stress, microbial community diversity and composition are primarily shaped by deterministic assembly processes [24,25]. The strong selective pressure from metal toxicity and altered soil properties favors taxa with specialized resistance traits, such as efflux pumps and detoxification enzymes [14,25,26].

In Bulgaria, heavy metals represent the primary pollutants affecting soil quality, due to intensive mining and smelting activities. The accumulation of HMs in agricultural soils, particularly in smelters' vicinities, has become a pressing environmental issue, posing serious threats to food safety and human health. In the scientific literature, taxonomic shifts in microbial communities inhabiting HM polluted soils near smelters are well documented; however, most research has focused on (i) bacteria, with comparatively limited characterization of fungal communities and (ii) relatively low levels of HM pollution. Moreover, knowledge about the distribution of key pathways of HM resistance in multi-metal polluted soils remains incomplete. To bridge this gap, this study aims (i) to assess the ecological risk of soils for their inhabitants; (ii) to evaluate bacterial and fungal diversity and the shifts in their composition and structure across the HM pollution gradient, and (iii) to predict key metabolic pathways associated with HM resistance in bacterial communities. To achieve these aims, 16S rRNA and Internal Transcribed Spacer (ITS) ribosomal RNA sequencing were employed. The study focused on the area near the Pb-Zn smelter "KCM 2000", located in the Plovdiv region of South-Central Bulgaria. Previous research has reported extremely high levels of soil polymetallic pollution (Pb, Zn, Cd, Cu, and As) in this region [27]. We hypothesized that HM pollution, as the primary factor, along with local soil properties, shaped the community structure and microbial distribution in long-term HM polluted soils, with key taxa maintaining stable and persistent profiles along the pollution gradient.

2. Materials and Methods

2.1. Study Area and Soil Sampling

The study area is located near the "KCM 2000" smelter (42°03′40.8″ N, 24°48′52.0″ E) (Figure 1) with more than 60 years of operational history. Long-term smelting activities have significantly impacted the local environment, particularly the soils, which are heavily polluted with Pb, Zn, Cd, Cu, and As [27].

Soil samples were collected in June 2020 from agricultural fields with industrial and oilseed crops (*Lavandula vera* L., *Gossypium* sp., *Brassica napus*). Five sampling sites were selected, depending on their distance from the source of pollution and in relation to previous data [27]. The sampling sites were as follows: KCM_1, located 0.5 km south of the smelter (42°03′31.68″ N, 24°49′19.2″ E); KCM_2, located 2 km south of the smelter (42°03′5.76″ N, 24°49′19.6″ E); KCM_3, located 3 km south of the smelter (42°02′6″ N, 24°49′19.2″ E);

KCM_4, located 1 km southeast of the smelter (42°03′31.68″ N, 24°49′45.12″ E); KCM_5, located 1 km northeast of the smelter (42°04′23.52″ N, 24°49′45.12″ E). Soils were sandy loam textured with slightly acidic (KCM 4.1 and KCM_5) to neutral (KCM_1, KCM_2, KCM_3, KCM_4.2) pH and total organic carbon ranged from 7.04 g/kg (KCM_5.1) to 16.07 g/kg (KCM_4.2).

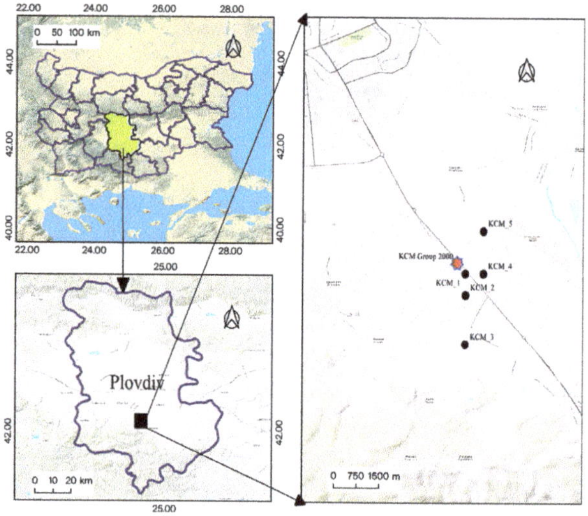

Figure 1. Map of sampling sites in the area of the "KCM 2000" smelter, Bulgaria.

Five subsamples (200 g) were randomly collected at each site within a 100 m × 100 m quadrat from two soil depths: surface (0–20 cm) and subsurface (20–40 cm) layers. Post-collection, the soil samples were sieved, and a representative sample from each site and soil depth was prepared by mixing the subsamples into aliquots. The representative samples were divided into two portions: one was stored at −80 °C for molecular analysis, and the other at 4 °C for HM analyses.

2.2. Soil Concentrations of Heavy Metals and Ecological Risk Index

Heavy metal content in soils was measured as described in [8], and the concentrations of HMs are shown in Table S1. The potential ecological risk index (ERI) was calculated for both total and bioavailable forms of HMs according to [28]. Soil toxicity risk was assessed using the scale proposed by the same authors: ERI < 100 (low risk), 100 ≤ ERI < 150 (moderate risk), 150 ≤ ERI < 200 (considerable risk), 200 ≤ ERI < 300 (very high risk), and ERI ≥ 300 (disastrous risk). Due to the lack of relevant background data on unpolluted soil in the study area, the average of values below the maximum permissible concentrations (MPCs) specified by Bulgarian Regulation 3/2008 (https://www.moew.government.bg/bg/pochvi/zakonodatelstvo/nacionalno-zakonodatelstvo/ (accessed on 1 August 2008)) for each HM was used as a reference. Since the soil levels of Pb and Cd exceeded the MPCs, the respective MPC values were used in calculating the ERIs.

2.3. DNA Extraction and Illumina Sequencing

Total DNA was extracted from 0.5 g of soil using the E.Z.N.A. DNA Soil Kit (Omega Bio-tek, Norcross, GA, USA) following the manufacturer's protocol. DNA concentration and purity were evaluated using a NanoDrop 1000 spectrophotometer (Thermo Scientific, Waltham, MA, USA) and verified by 1% agarose gel electrophoresis. Quantification was performed with a Qubit 4 Fluorometer (Thermo Scientific, Waltham, MA, USA).

For bacterial community analysis, the hypervariable V3–V4 regions of the 16S rRNA gene were amplified using barcoded primers 341F (5′-CCTACGGGNGGCWGCAG-3′) and 806R (5′-GACTACHVGGGTATCTAATCC-3′) [29]. Amplicon libraries were prepared and sequenced on an Illumina MiSeq platform (Macrogen, Seoul, Republic of Korea) with a read length of 2 × 300 bp. Fungal ITS2 regions were amplified and sequenced following the methodology described by [30] in the Laboratory of Environmental Microbiology, Institute of Microbiology, Czech Academy of Sciences, Prague. This included PCR amplification with primers gITS7 (5′-GTGAATCATCGAATCTTTG-3′) and ITS4 (5′-TCCTCCGCTTATTGATATGC-3′), library preparation using the TruSeq PCR-Free Kit, (Illumina, San Diego, CA, USA) and sequencing on an Illumina MiSeq platform with a read length of 2 × 250 bp.

2.4. Bioinformatics and Data Processing

After trimming the barcodes, the raw sequencing data were processed using the standard QIIME2 pipeline [31]. Paired-end reads were denoised with DADA2 [32], applying truncation lengths of 225 and 300 bases for bacterial data and 173 and 131 bases for fungal data, based on read quality plots. Predicted functional changes in the bacterial community were assessed using the Phylogenetic Investigation of Communities by Reconstruction of Unobserved States (PICRUSt2) software package [33]. The complete PICRUSt2 pipeline, as implemented in the QIIME2 plug-in, was run on the denoised paired-end reads. The KO IDs obtained were manually annotated using the KEGG database to estimate HM resistance gene abundances.

Phylogenetic diversity analysis utilized trees generated with the "mafft" alignment method within QIIME2. Diversity metrics and group significance tests were calculated using the "diversity" plug-in library. The "core-metrics-phylogenetic" method was applied with a sampling depth of 12,650, determined by inspecting the feature table's interactive sample detail. Taxonomic classification of bacterial sequences was performed using a Naive Bayes classifier pre-trained on the Greengenes 13_8 99% OTU reference database (2024.09.taxonomy) [34–36]. For fungal sequences, taxonomic classification was conducted using the UNITE 9 database [37]. OTUs were clustered at a 99% sequence identity threshold. Differential abundance testing between groups was performed using the Analysis of Composition of Microbiomes (ANCOM) method, implemented via the "q2-composition" plug-in, with pseudo-counts added to address zero frequencies. Chimera checking was conducted on aligned 16S rRNA and ITS sequences using the UCHIME algorithm [38].

2.5. Data Analyses

One-way ANOVA, followed by Tukey's test, was performed to examine the significance of the differences in biotic and abiotic parameters.

The similarity of microbial communities was evaluated using a multi-dimensional scaling (NMDS) technique and taxonomic data on bacterial and fungal classes. NMDS was also used to visualize the distribution of KEGG pathways.

Two-way PERMANOVA, followed by SIMPER analyses, were conducted to assess the impact of ERI and soil depth on microbial distribution and identify the microbial classes most strongly contributing to intergroup differences in overall microbial community dissimilarity.

Pearson correlation analysis was performed to assess the relationships between ERI/soil depth, microbial classes and KEGG distributions.

Alpha diversity indices (Chao1, Shannon) were calculated based on bacterial and fungal taxonomic data.

The above statistics were performed with the PAST package [39] at a significance level of $p < 0.05$.

3. Results

3.1. Soil Concentrations of HMs and Ecological Risk Index

The HM pollution was severe, with mean soil concentrations recorded across soils as follows: As, 46.03 mg/kg; Cd, 46.87 mg/kg; Cu, 272.84 mg/kg; Pb, 2824 mg/kg; and Zn, 2991 mg/kg (Table S1). In most cases, these concentrations exceeded the MPC. In the context of total HM concentrations, ERIs were calculated (Table 1), and a gradient of HM ecological risk was established: Cd (disastrous) >> Pb (very high) >> As (moderate) >> Cu (low) > Zn (low).

Table 1. Ecological risk indices (ERIs) were calculated per HM and sampling site (total ERI).

Soil	Zn	Pb	Cd	Cu	As	Total ERI
KCM_1.1	40.55	578.45	2773.50	59.41	184.36	3636.27
KCM_2.1	6.68	68.51	235.50	9.26	12.93	332.88
KCM_3.1	0.93	6.78	58.50	3.58	9.27	79.05
KCM_4.1	29.48	286.15	1293.00	38.01	77.22	1723.86
KCM_5.1	3.17	16.75	139.50	6.54	14.96	180.93
KCM_1.2	14.83	187.20	927.00	22.07	58.49	1209.59
KCM_2.2	7.34	74.02	579.00	10.34	4.83	675.52
KCM_3.2	1.07	6.50	54.00	4.00	8.69	74.26
KCM_4.2	20.78	171.20	835.50	22.76	61.29	1111.53
KCM_5.2	3.47	16.50	135.00	5.88	12.26	173.11

The gradient of polymetallic ERI was arranged from the lowest to the highest risk as follows: KCM_3.2 < KCM_3.1 (low) << KCM_5.2 < KCM_5.1 (considerable) << KCM_2.1 < KCM_2.2 < KCM_4.2 < KCM_1.2 < KCM_4.1 < KCM_1.1 (disastrous). The difference in ERI between the surface- and subsurface soil layers was insignificant ($F = 1.07$; $p = 0.47$).

Additionally, bioavailable forms of Pb, Zn and Cd were measured in the surface soil layer, and their amounts varied, on average, from 0.03–0.08% (Pb and Zn) to 4.78% (Cd) from the respective total HM concentration (Table S2). The gradient of soil toxicity based on bioavailable forms of HMs arranged from the lowest to the highest ERI was as follows: KCM_2.1 << KCM_5.1 < KCM_3.1 << KCM_4.1 << KCM_1.1 (Table S2).

3.2. Taxonomic Composition and Structure of Soil Bacterial Communities

Bacterial communities consisted of 15 phyla (Figure 2A). The dominant phylum was Proteobacteria, followed by Actinobacteriota and Acidobacteriota, which accounted for 26.96 ± 6.2%, 18.55 ± 5.8%, and 14.30 ± 4.1% of the total bacterial community, respectively. The subdominant phyla in soils were Bacteroidota, Chloroflexota, Gemmatimonadota and Planctomycetota, ranging from 4.87% to 7.06% of the total bacterial communities. Notably, Gemmatimonadota and Planctomycetota revealed a strong positive relationship with ERI ($r \geq 0.84$; $p = 0.002$), representing up to 9.84% in the KCM_1.1 and up to 8.47% in the KCM_1.2 bacterial communities. Other predominantly local distributions were observed for Verrucomicrobiota and Myxococcota_A_473307 (KCM_2), Firmicutes_D (KCM_5), and Nitrospirota_A_437815 (KCM 1 and KCM_2). Low-abundant phyla Firmicutes_A and Methylomirabilota were represented in KCM_5.2 and all subsurface soils (except KCM_2.2), respectively. The effect of depth on the distribution of bacterial phyla was insignificant.

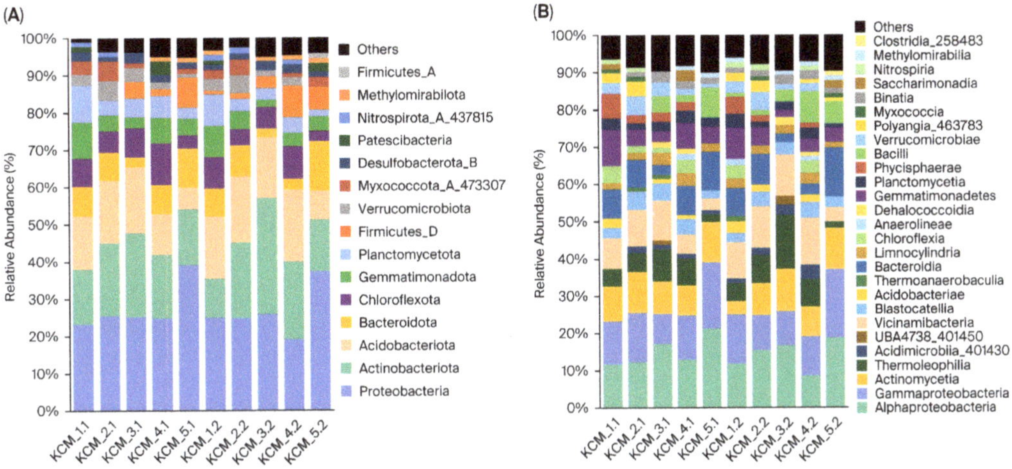

Figure 2. Relative abundance of bacterial phyla (**A**) and bacterial classes (**B**) in KCM soils. 'Others' represents a sum of relative abundances of phyla/classes < 1%.

At the class level, taxonomic profiling of bacterial communities identified 27 classes (Figure 2B). The most widespread and dominant class across all soils was Alphaproteobacteria, averaging 14.66 ± 3.82%, followed by Gammaproteobacteria (12.30 ± 3.50%) and Actinomycetia (9.14 ± 2.33%). Subdominant classes in the bacterial community composition were Vicinamibacteria (8.53 ± 3.36%), Bacteroidia (6.99 ± 3.50%) and Thermoleophilia (6.35 ± 3.68%). Depth dependence, although insignificant in class distribution, was observed for Methylomirabilia, Acidimicrobiia_401430, Vicinamibacteria and Clostridia_258483 ($r < 0.39$; $p > 0.26$).

The diversity of bacteria at the class level (Table S3) was estimated by the following: (1) bacterial richness (Chao1), with a slightly, but not significantly higher number in the surface soil layer (492 ± 40) compared to the subsurface one (432 ± 90); (2) Shannon diversity index, with a higher but insignificantly different mean value in surface soils. The two indices showed significant correlation ($F = 0.97$, $p < 0.0001$), indicating the major role of taxonomic heterogeneity (Chao1) instead of class evenness in determining Shannon index values.

3.3. Taxonomic Composition and Structure of Soil Fungal Communities

The fungal communities were represented by 4 phyla (Figure 3A), with the dominance of Ascomycota, accounting for 95.20–99.79% of the fungal reads. Basidiomycota exhibited a local distribution, occurring in low abundances (<1.30%) in KCM_3. and KCM_4.2.

At the class level, 10 distinct fungal classes were identified in the studied soils (Figure 3B). Among these, Sordariomycetes, Eurotiomycetes and Dothideomycetes were almost universally distributed, except Dothideomycetes in KCM_5.1. Sordariomycetes and Eurotiomycetes can be considered dominant classes in fungal communities. The other two classes were considered as subdominants. A significant effect of soil depth was not recorded.

Fungal diversity across soils was estimated by Chao1 richness, ranging from 44 (KCM_1.1) to 118 (KCM_4.1), and Shannon index, ranging from 3.78 (KCM_1.1) to 5.99 (KCM_2.1) (Table S3). The surface soil layer showed higher values than the subsurface one, but the differences were insignificant ($F \leq 2.09$; $p \geq 0.15$).

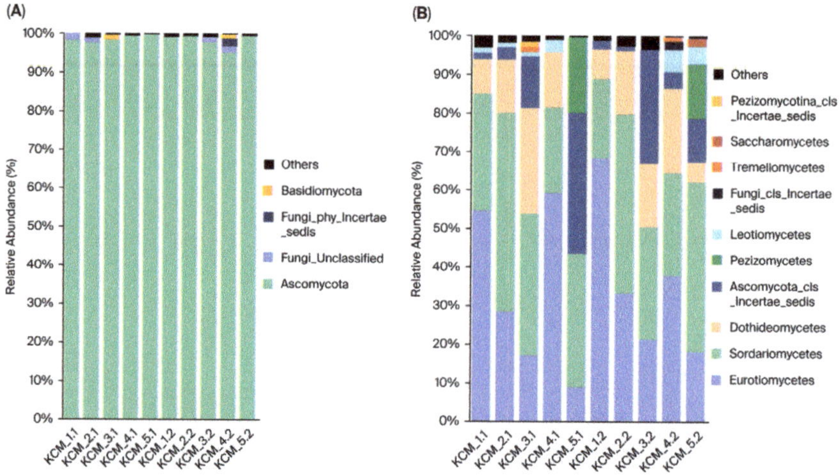

Figure 3. Relative abundance of (**A**) fungal phyla and (**B**) fungal classes in KCM soils. 'Others' represents a sum of relative abundances of phyla/classes < 1%.

3.4. Microbial Soil Similarity

Non-metric multidimensional scaling (NMDS) was used to visualize the spatial arrangement of soils in ordination space, where the distances between points represented the class similarity of microbial communities (Figure 4). The clustering pattern of soils based on bacterial (Figure 4A) and fungal (Figure 4B) community compositions segregated the soils into two distinct groups: green colored with ERIs $\leq$ 180, and brown colored with ERIs $\geq$ 333. In both ordination plots, no overlapping of convex hulls was observed, indicating taxonomic differences between the soils. The NMDS analysis also revealed that some bacterial and fungal classes were distributed only in the convex hull of soils with low to considerable ecological risk (UBA4738_401450, Clostridia_258483, and Pezizomycetes), while others (Verrucomicrobiae, Thermoanaerobaculia, Chloroflexia and Eurotiomycetes) were associated with disastrous risk soils. A third group of microbial classes showed a more random distribution pattern, indicating a high tolerance to environmental factors, including HMs.

To evaluate the impact of potential soil toxicity (ERIs) and soil depth on microbial (bacterial and fungal) distribution, a two-way PERMANOVA was conducted (Table S4). PERMANOVA validated the NMDS ordination results, confirming the significant effect of soil toxicity ($F \geq 4.97$; $p < 0.002$) on microbial distribution. In contrast, the effects of soil depth ($F \leq 0.69$; $p \geq 0.62$), and the interaction between soil depth and soil toxicity risk ($F \leq 0.82$; $p \geq 0.48$) were insignificant.

To identify the classes most strongly contributing to the differences in overall bacterial/fungal community dissimilarities, SIMPER analyses were conducted (Tables S5 and S6). The overall dissimilarity among bacterial communities was 31.79%, with six classes accounting for approximately 53% of this variation (Table S5). These classes were Alphaproteobacteria, Thermoleophilia, Gammaproteobacteria, Bacteroidia, Vicinamibacteria and Bacilli. Among these classes, Alphaproteobacteria ($r = -0.57$) and Bacilli ($r = -0.36$) showed strong but insignificant correlations with ERIs, while the relationships between other classes and soil toxicity risk were weak ($r < -0.16$) (Table S7). The overall average dissimilarity among fungal communities was 35.32% (Table S6) and Eurotiomycetes contributed most to this dissimilarity. Eurotiomycetes were associated with soils, having an ERI $\geq$ 333. Pearson correlation analysis confirmed strong correlations between ERI and the distribution of the class Eurotiomycetes (r = 0.73; $p = 0.016$) (Table S8).

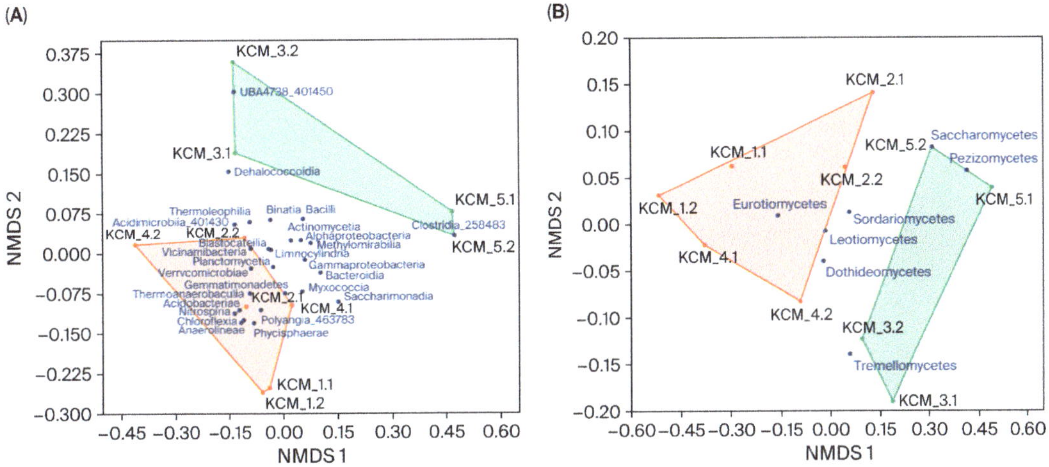

Figure 4. Non-metric multidimensional scaling (NMDS) plot of soil taxonomic similarity based on Euclidean resemblance matrix, calculated on (**A**) bacterial and (**B**) fungal classes. Two-dimensional stress: (**A**) 0.079 and (**B**) 0.088. Convex hulls of soils with ecological risk index (ERI) $\leq$ 181 and ERI $\geq$ 333 are shown in green and brown, respectively.

3.5. Predicted Bacterial Resistance to Heavy Metals

Using the KEGG database, the functional pathways related to HM resistance of bacteria were identified among the total observed pathways (Table S9). A total of 26 dominant HM resistance pathways were selected, with a cumulative frequency of 717680. KEGG abundance was highest in KCM_1 (averaging 12.5%) and lowest in KCM_3 (averaging 7.9%). The most abundant pathway was K15726, followed by K07240 and K16264 (Table S10). Pathway K16264, associated with the genes *czc*D and *zit*B, which confer Zn-, Co-, and Cd-resistance, occurred at relatively high abundance across all soils. Genes involved in the HM efflux system czcA, *cus*A, and *cnr*A (K15726) were highly abundant in the severely polluted soils KCM_1 and KCM_2, and least abundant in the low-polluted soil KCM_3.2. All of these genes encode cation diffusion facilitators and efflux systems that transport bivalent metal ions out of the cytoplasm. Chromate efflux and reduction pathways K07240 and K19784, regulated by *chr*A and *chr*R, were detected in all soils, with K07240 showing high abundance and K19784 relatively low abundance. The Cu/Ag efflux pathways K07787 (*cus*A, *sil*A) and K07798 (*cus*B, *sil*B) were common across all soils, particularly in KCM_1 and KCM_2.1. The least abundant pathway, K12951 (*ctp*D), related to Co/Ni transport, was detected only in KCM_1. Additionally, less-distributed pathways were identified in KCM_3, including K00520, K08363, and K19058, all involved in bacterial resistance to Hg.

To arrange the KEGG distribution across soils, NMDS was applied (Figure 5). NMDS technique ordinated 16 of 26 KEGG pathways in the convex hull of soils exhibiting disastrous risk (ERI $\geq$ 333), and none in the convex hull of soils with ERI $\leq$ 181. Some of the pathways were located outside of the two convex hulls, suggesting a relatively uniform distribution across soils.

Pearson correlation analysis confirmed the NMDS results, showing significant or insignificant but strong relationships between ERIs and KEGGs distribution (Table S10). Most correlations were positive, except for K05792, K03893, and K11811, which are pathways associated with bacterial resistance to the metalloids As and Te. The distribution of

KEGGs across soil depth was insignificant and predominantly negative, although several relationships (K16264, K19784, K07156, and K07245) were notably strong.

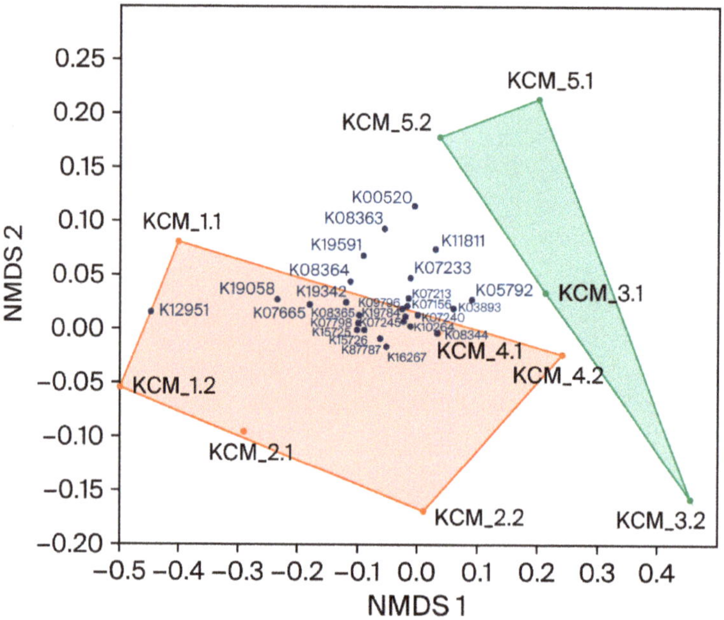

Figure 5. Non-metric multidimensional scaling (NMDS) plot of Kyoto Encyclopedia of Genes and Genomes (KEGG) pathways based on Euclidean resemblance matrix, calculated on bacterial 16S rRNA sequence data. Two-dimensional stress: 0.052. Convex hulls of soils with ecological risk index (ERI) $\leq$ 181 and ERI $\geq$ 333 are shown in green and brown, respectively.

4. Discussion

4.1. Potential Toxic Risk of Soil Environments

Soils in the vicinity of the "KCM 2000" smelter have been subject to HM pollution since the plant's early operation. In the present study, the ecological risk index calculated from total HM concentrations ranged from low (KCM_3) to considerable (KCM_5) and disastrous (KCM_1, KCM_2, and KCM_4). Soils classified within the disastrous risk category exceeded the lower threshold of this range [28] by factors of two (KCM_2.2) to ten (KCM_1.1). Based on these results, the studied soils were categorized into two groups: (i) those with HM levels potentially tolerable by soil microorganisms without substantial alterations in community structure (KCM_3 and KCM_5), and (ii) those with pollution levels likely to induce significant shifts in microbial community composition and function (KCM_1, KCM_2, and KCM_4).

Additionally, the ecological risk was calculated based on bioavailable forms of HMs, and the values of ERIs classified the soils as low risk (KCM_2.1, KCM_3.1 and KCM_5.1), at the boundary between considerable and very high risk (KCM_4.1), and disastrous risk (KCM_1.1). It was reported [40] that metal toxicity is more closely related to the concentration of bioavailable forms than to total metal concentrations. On the other hand, the concentrations of bioavailable forms of HMs are much more variable than the total metal content and depend on metal speciation and the nature of its salts, as well as climatic and soil conditions [41,42]. Therefore, the bioavailable metal concentrations reported in the study are considered a snapshot of their possible soil levels, accounting for less than 0.24% of the total metal content, except for Cd, which ranges from 1.3% to 12.8%.

Cd and Pb in both forms (total and bioavailable) contributed the most to the calculated ERI values, owing to their high concentrations (bioavailable Cd and total Pb) and environmental toxicity (Cd) [21,43]. The average values of Cd's single-factor environmental risk indices were 703 (total content) and 336 (bioavailable content), both of which fall under the category of disastrous environmental risk, making Cd the primary contributing factor to soil toxicity. In contrast, the single-factor environmental risk index for Pb was 141 for total content and 24 for bioavailable content, placing Pb in two distinct risk categories very high and low, respectively, depending on its chemical form. The low bioavailable risk level is attributed to Pb's low water solubility [44], particularly in soils with neutral pH [45], which limits its mobility and microbial toxicity. Therefore, maintaining stable soil conditions, especially preventing acidification, is essential to avoid increased Pb solubilization and the consequent rise in its environmental risk.

4.2. Microbial Distribution

Long-term heavy metal (HM) pollution disrupts ecological balance and alters microbial community composition and function. This impact is particularly evident near "KCM 2000"—Plovdiv, where HM pollution remains severe and persistent. However, information on microbial community responses to such stress in this area is still limited. This study, therefore, examines taxonomic shifts in bacterial and fungal communities across different HM pollution levels and soil depths. Distribution of bacterial and fungal phyla across KCM soils harbors 15 bacterial and 4 fungal phyla. We assume that the high difference in the presence of phyla from the two microbial domains is largely related to inherent life strategies rather than higher fungal sensitivity to HMs. According to [46], bacteria, as r-strategists, can be represented by up to 8800 different species genomes, depending on soil type, while soil fungi, as K-strategists, form less diverse communities, occupying broader ecological niches (particularly trophic), and dominating in soil microhabitats in terms of biomass [47].

The bacterial communities were dominated by Proteobacteria, Actinobacteriota, and Acidobacteriota (Figure 2A), which were consistently identified as major phyla across all studied soils. Previous studies have similarly reported Proteobacteria, Acidobacteria, Actinobacteriota, and Chloroflexota as dominant taxa in smelting-polluted environments [48–51], while Firmicutes has been highlighted as one of the most resistant phylum in such environments [49]. The dominant phyla were also detected in both unpolluted soils [52,53] and HM polluted soils across various land uses, including agricultural fields [6,17,50,54] and areas affected by mining and industrial effluents [55,56]. Their widespread presence and persistence underscore their inherent tolerance and adaptability to HM stress, making them crucial components of microbial resilience in polluted ecosystems. Planctomycetota and Gemmatimonadota demonstrated higher relative abundance specifically for highly polluted soils KCM_1. It was also reported Planctomycetota and Gemmatimonadota are resistant to HMs and can survive in HM polluted soils [17].

Fungal KCM communities were dominated by Ascomycota. Several authors [57–59] have reported that Ascomycota colonize a wide range of environments, both disturbed and undisturbed. Moreover, Ascomycota was the highest relatively abundant fungus in HM polluted soils [17]. This adaptability of Ascomycota is attributed to a high number of genes associated with nutrition, carbohydrate metabolism, and a significantly high frequency of genomic traits linked to both stress tolerance and competitive abilities [57].

4.3. Distribution of Bacterial and Fungal Classes Across Soils

At the class level, bacterial richness and diversity were high and showed no significant differences between soils and soil depths. The equal bacterial distribution suggested that adaptation processes have occurred, likely selecting highly environmentally tolerant and flexible bacterial forms. The three most widespread bacterial phyla in KCM soils were represented by 10 of 27 identified classes. Alphaproteobacteria, Gammaproteobacteria, and Actinomycetia were identified as the dominant classes. High abundances of Alphaproteobacteria and Gammaproteobacteria have also been reported in smelting-polluted soils, along with Betaproteobacteria and Gemmatimonadetes [60,61]. Additionally, Thermoleophilia and Vicinamibacteria, Blastocatellia (Acidobacteriota), Bacteroidia (Bacteroidota), and Planctomycetia (Planctomycetota) were also relatively highly represented across soils.

All these classes were localized at an ordination plot (NMDS) between soil groups with low to considerable (KCM_3 and KCM_5) and disastrous (KCM_1, KCM_2, and KCM_4) ecological risks, and determined by SIMPER analysis as the main contributors to soil bacterial dissimilarity. Pearson correlation analysis confirmed the NMDS results, revealing weak relationships (either negative or positive) between bacteria from the aforementioned classes and ERIs, except for Alphaproteobacteria and Bacilli. The last two bacterial classes related strongly (although not statistically significant) and negatively to soil ecological risk, suggesting their preference to low-polluted soil habitats. For instance, [62] also found a reduction in Alphaproteobacteria abundance in agricultural soil polluted with Cd (1639 mg/ kg). In contrast, *Bacillus* sp. (class Bacilli) was highly abundant in Cd-polluted soils and could have potential for bioremediation in such sites [60]. Alphaproteobacteria and Bacilli were observed by other authors [63–65] as the most soil-abundant classes, which harbor a wide range of habitats, various kinds of metabolic processes, and cellular phenotypes. It includes mostly phototrophic genera, plant and animal pathogens, symbiotic relations with plants, phytohormone producers, etc.

Further, NMDS ordination positioned classes Chloroflexia (Chloroflexota), Gemmatimonadetes (Gemmatimonadota), Verrucomicrobiae (Verrucomicrobiota), Phycisphaerae (Planctomycetota), and Acidobacteriae (Acidobacteriota) within the convex hull of soils with disastrous ecological risk. All of these classes, except Gemmatimonadetes, showed significant positive correlations with ERIs and contributed approximately 4% to the inter-soil bacterial dissimilarity. Notably, the relative abundance of these classes increased in the most at-risk soil (KCM_1) compared to their mean abundances in the other soils. The increased abundances of certain bacterial taxa in highly HM-polluted soils have been reported by many authors [4,5,62], with these shifts in bacterial communities being attributed to the selectivity of HMs. The abundance of Verrucomicrobiae in KCM_1 was the same as the mean abundance for other soils. Based on the relatively even distribution of Verrucomicrobiae across soils and its negative correlation with ERI, we propose that this bacterial class should be classified alongside Alphaproteobacteria, Gammaproteobacteria, etc., rather than with the classes detected within the convex hull of disaster-risk soils. This class likely occupies a boundary position, as indicated in the ordination plot, where it is clearly situated between soils KCM_2.1 and KCM_2.2, which possessed the lowest ERI values compared to other soils in the disaster-risk category.

Finally, the distribution of certain bacterial classes (Acidimicrobiia_401430, Thermoanaerobaculia, and Binatia) in most soils, except for KCM_1 (ERI: 3636), suggests that representatives of these classes are generally HM-tolerant but not to the extreme pollution levels observed in KCM_1. On the contrary, the class UBA4738_401450, which was detected only in KCM_3 (low ecological risk), could be considered sensitive to HMs. However, the Pearson correlation analysis did not confirm these trends of distribution. All these bacterial classes correlated negatively with soil ecological risk, with the correlation for Acidimicrobiia_401430 ($ERI_{bioavailable\ HM}$) and Binatia (both $ERI_{total\ HM}$ and $ERI_{bioavailable\ HM}$) being statistically significant. In this context, we assumed that soil toxicity non-linearly influenced bacterial distribution; likely this effect was modulated by local soil properties [66]. Our previous study [8] demonstrated the influence of soil pH, soil texture, and total organic carbon, along with HMs, on bacterial abundance and distribution in KCM soils. In general, in the context of observed trends in bacterial distribution, we propose the presence of three main groups of bacteria concerning soil toxicity and environmental stress: resistant (Chloroflexia, Gemmatimonadetes, Phycisphaerae and Acidobacteriae), tolerant (Alphaproteobacteria, Gammaproteobacteria, Actinomycetia, Verrucomicrobiae and many others), and sensitive (UBA4738_401450). The primary driving factors of their distribution were HM concentrations and local soil properties, but not soil depth. The Pearson correlations between bacterial distribution and soil depth were insignificant, with most being weak and negative.

The fungal diversity in KCM soils was represented by 10 classes, eight of which belong to the phylum Ascomycota. Fungal taxonomic richness and diversity were relatively lower compared to bacteria, but very similar to that shown in other studies [15]. The dominant classes in these soils included Eurotiomycetes and Sordariomycetes, while Dothideomycetes were classified as sub-dominants. Additionally, Pezizomycetes and Leotiomycetes (Ascomycota) were also widely distributed across the studied soils, but at low relative abundance. Tremellomycetes (Basidiomycota) were observed in very low abundances and showed localized distribution in KCM_2 and KCM_3, respectively.

NMDS ordination segregated the fungal classes Eurotiomycetes and Leotiomycetes within the convex hull of disastrous-risk soils, classifying them as resistant to HMs. Pearson correlation analysis further confirmed the positive correlations of these classes, particularly Eurotiomycetes, with the levels of soil toxicity. For example, Eurotiomycetes were present in low-risk soils of KCM_3 at 19.52%, in considerable-risk soils of KCM_5 at 13.64%, and in disastrous-risk soils at 46.44%. The resistance of Eurotiomycetes to HMs has been documented in early studies, where the authors demonstrated that their relative abundance in community structure increased with higher Cu and Cd [67], and Cd, Cr, Sn and Hg [68] pollution levels. Members of this class, such as *Aspergillus niger* and *Penicillium* sp., appear to evolve specific catabolic activities to use many pollutants as nutrients and energy sources [69]. For example, *Aspergillus* species show strong resistance to Cr and Pb through bioaccumulation and extracellular enzyme production, making them promising organisms for bioremediation [70].

The other groups of fungal classes segregated on the ordination plot included tolerant (Sordariomycetes, Dothideomycetes, and Tremellomycetes) and sensitive (Pezizomycetes and Saccharomycetes) classes. The abundances of these classes related negatively to HMs, which, although mostly strong, were statistically insignificant. The classes Pezizomycetes and Saccharomycetes had localized distribution in KCM_5. This distribution of Pezizomycetes confirmed the findings of [71] about the higher abundance of this class in soils with low concentrations of HMs. The class Sordariomycetes was evenly distributed across soils with varying ERI levels, ranging from 33.10% in low-risk soils to 39.01% in disastrous-risk soils. However, its highest relative abundance was observed in KCM_2 (average ERI: 504) and the lowest in KCM_1 (average ERI: 2422). This suggests that members of Sordariomycetes are more tolerant rather than resistant to HMs. Moreover, *Fusarium* species, which belong to this class, show strong biosorption and biotransformation abilities in Zn sequestration through organic acid production, intracellular compartmentalization, and enzymatic detoxification [70].

PERMANOVA identified HM concentrations as a driving factor of bacterial and fungal class segregation.

4.4. Distribution of HM Resistance Genes and Predicted Bacterial Resistance Pathways

In this study, the taxonomic distribution of bacteria in soils with varying ecological risk levels was analyzed alongside the distribution of genes encoding HM resistance. The results showed that pollution increased the frequency of HM resistance genes, with the highest frequency observed in the most ecological risk soils. This finding confirmed the results of earlier studies dedicated to HM toxicity and the distribution of potential functional pathways for resistance [72,73]. NMDS ordination revealed that approximately 62% of the dominant KEGG pathways were localized in disastrous-risk soils, demonstrating their role in adaptation to HM stress and resistance of bacteria. This was supported by the predominantly positive correlations between KEGG pathway abundances and ERIs. Among the 16 pathways linked to severe HM pollution, the most abundant were K12951, K15726, K15725, K07665, K07787, and K07798, which were associated with Cu, Co/Ni, and Cr efflux systems, which, after modification, could also be utilized by bacteria for defense against Zn, Cd, and Pb [74]. We suggest that KEGG pathways are related to the relatively abundance of Proteobacteria and classes Alphaproteobacteria and Gammaproteobacteria, which exhibit resistance to Cd and play important roles in Cd-contaminated soils [60,70]. The mechanism underlying Cd resistance is associated with the precipitation–dissolution balance, which constrains the dynamic fluctuations of free metal ions in the soil [75]. Our findings align with [62], who reported enrichment of such metabolic pathways in Cd-contaminated soils, indicating that microbial communities modify their metabolic processes to mitigate HM stress.

KEGG pathways positioned between the convex hulls on the NMDS plot were assigned with wider distribution and higher frequencies in lower- to considerable-risk soils compared to disastrous-risk soils. The primary defense mechanisms of these pathways for HM detoxification include enzyme activities, synthesis of periplasmic binding proteins, ion transport and ion efflux systems. Some of these defense pathways are widespread in protecting cells against excessive HM concentrations. It has been shown that HM resistance is the result of multiple layers of resistance systems with overlapping substrate specificities, but unique functions [74]. According to [74], some of these systems provide basic cellular defense against HMs, while others are highly specialized and found only in a few bacteria, conferring specific HM resistance. In general, predictive profiling identified efflux systems and detoxifying enzymes as prominent mechanisms for HM resistance (Tables S9 and S10). Among the metabolic pathways, those related to reducing the toxicity of Cr, Zn and Cd were most abundant, highlighting microbial ability to thrive in toxic environments by actively removing or immobilizing these metals.

5. Conclusions

The agricultural soils near the "KCM-2000" smelter were identified as risky from low to disastrous levels due to HM pollution, with Cd identified as the major contributor to toxicity. Microbial distribution patterns indicated the presence of HM-tolerant and HM-resistant bacterial and fungal phyla/classes. A strong positive correlation was observed between ERIs and the distribution of KEGGs, highlighting the adaptation that occurred in microbial communities according to the newly created soil environments. KEGG functional analysis revealed both widespread among soil locations common HM defense pathways and more complex mechanisms for HM-resistant bacteria in highly polluted soils. These results will be further validated in forthcoming studies. Overall, the findings highlight the substantial influence of the smelting industry on soil microbial diversity in surrounding areas and underscore the potential of microbially based approaches for monitoring soil health and promoting safe agricultural practices.

Supplementary Materials: The following supporting information can be downloaded at https://www.mdpi.com/article/10.3390/toxics13121025/s1, Table S1: Total concentrations of heavy metals/metalloids (HMs) in the surface and subsurface soils; Table S2: Water soluble concentrations of heavy metals (Pb, Zn, and Cd) in the surface soils, and their individual and total ecological risk index (ERI); Table S3: Diversity estimation statistics of microbial communities; Table S4: Results of the two-way PERMANOVA indicating the effects of soil toxicity (expressed as the Ecological Risk Index, ERI) and soil depth on the distribution of bacterial classes; Table S5: SIMPER results showing the contribution of bacterial classes to the total dissimilarity (31.79%) among soil bacterial communities; Table S6: SIMPER results showing the contribution of fungal classes to the total dissimilarity (42.03%) among soil fungal communities; Table S7: Pearson correlation coefficients between the relative abundances of bacterial classes, and soil toxicity (ERI) and depth (n = 10 for total HMs; n = 5 for bioavailable HMs). The level of significance is indicated in brackets. Significant correlations are bolted; Table S8: Pearson correlation coefficients between the relative abundances of bacterial classes, and soil toxicity (ERI) and depth (n = 10 for total HMs; n = 5 for bioavailable HMs). The level of significance is indicated in brackets. Significant correlations are bolted; Table S9: KEGG pathways providing bacterial heavy metal resistance in soils; Table S10: Definition of the KEGG pathways providing heavy metal resistance of bacteria in soils; Table S11: Pearson correlation analysis between the relative KEGG abundances, and soil risk of toxicity (ERI) and depth (n = 10 for total HMs; n = 5 for bioavailable HMs). The level of significance is indicated in brackets. Significant correlations are bolted.

Author Contributions: R.N. Investigation, Methodology, Formal analysis, Writing—original draft, E.G.: Formal analysis, Validation, Data curation; A.K. Investigation, Writing—review and editing, Data curation, Formal analysis, Validation; M.P. Formal analysis, Validation; N.D. Formal analysis, Data curation, P.B. Investigation, Methodology, Data curation, Writing—review and editing, Funding acquisition, G.R. Writing—review and editing, Methodology, Conceptualization, Supervision. All authors have read and agreed to the published version of the manuscript.

Funding: This research was funded by the Ministry of Education, Youth and Sports of the Czech Republic, grant numbers CZ.02.01.01/00/22008/0004597.

Institutional Review Board Statement: Not applicable.

Informed Consent Statement: Not applicable.

Data Availability Statement: The datasets generated and analyzed during the current study are available in the NCBI Bioproject database under accession number PRJNA1363107.

Acknowledgments: The authors would like to thank the members of the Laboratory of Environmental Microbiology, Institute of Microbiology, CAS, for their support throughout this project.

Conflicts of Interest: The authors declare that they have no conflicts of interest.

References

1. Nyiramigisha, P.; Komariah; Sajidan. Harmful impacts of heavy metal contamination in the soil and crops grown around dumpsites. *Rev. Agric. Sci.* **2021**, *9*, 271–282. [CrossRef]
2. Campillo-Cora, C.; Rodríguez-Seijo, A.; Pérez-Rodríguez, P.; Fernández-Calviño, D.; Santás-Miguel, V. Effect of heavy metal pollution on soil microorganisms: Influence of soil physicochemical properties. A systematic review. *Eur. J. Soil Biol.* **2025**, *124*, 103706. [CrossRef]
3. Upadhyay, V.; Kumari, A.; Kumar, S. From soil to health hazards: Heavy metals contamination in northern India and health risk assessment. *Chemosphere* **2024**, *354*, 141697. [CrossRef] [PubMed]
4. Chodak, M.; Gołębiewski, M.; Morawska-Płoskonka, J.; Kuduk, K.; Niklińska, M. Diversity of microorganisms from forest soils differently polluted with heavy metals. *Appl. Soil Ecol.* **2013**, *64*, 7–14. [CrossRef]
5. Zampieri, B.D.B.; Pinto, A.B.; Schultz, L.; de Oliveira, M.A.; de Oliveira, A.J.F.C. Diversity and distribution of heavy metal-resistant bacteria in polluted sediments of the Araça Bay, São Sebastião (SP), and the relationship between heavy metals and organic matter concentrations. *Microb. Ecol.* **2016**, *72*, 582–594. [CrossRef] [PubMed]

6. Cui, H.; Liu, L.L.; Dai, J.R.; Yu, X.N.; Guo, X.; Yi, S.J.; Zhou, D.Y.; Guo, W.H.; Du, N. Bacterial community shaped by heavy metals and contributing to health risks in cornfields. *Ecotoxicol. Environ. Saf.* **2018**, *166*, 259–269. [CrossRef]
7. Hao, J.; Chai, Y.N.; Lopes, L.D.; Ordóñez, R.A.; Wright, E.E.; Archontoulis, S.; Schachtman, D.P. The effects of soil depth on the structure of microbial communities in agricultural soils in Iowa (United States). *Appl. Environ. Microbiol.* **2021**, *87*, 4. [CrossRef]
8. Nikolova, R.; Petkova, M.; Dinev, N.; Kenarova, A.; Boteva, S.; Berov, D.; Radeva, G. Correlation between bacterial abundance, soil properties and heavy metal contamination in the area of non-ferrous metal processing plant, Southern Bulgaria. *BioRisk* **2022**, *17*, 19–30. [CrossRef]
9. Nikolova, R.; Kenarova, A.; Boteva, S.; Dinev, N.; Radeva, G. Diversity and structure of soil bacterial communities in the area of non-ferrous metal plant revealed by 16S rRNA gene retrieval. *C. R. Acad. Bulg. Sci.* **2024**, *77*, 1260–1268. [CrossRef]
10. Coenen, A.R.; Hu, S.K.; Luo, E.; Muratore, D.; Weitz, J.S. A Primer for Microbiome Time-Series Analysis. *Front. Genet.* **2020**, *11*, 310. [CrossRef] [PubMed]
11. Meng, D.; Li, J.; Liu, T.; Liu, Y.; Yan, M.; Hu, J.; Li, X.; Liu, X.; Liang, Y.; Liu, H.; et al. Effects of redox potential on soil cadmium solubility: Insight into microbial community. *J. Environ. Sci.* **2019**, *75*, 224–232. [CrossRef] [PubMed]
12. Li, J.; Ma, Y.B.; Hu, H.W.; Wang, J.T.; Liu, Y.R.; He, J.Z. Field-based evidence for consistent responses of bacterial communities to copper contamination in two contrasting agricultural soils. *Front. Microbiol.* **2015**, *6*, 31. [CrossRef]
13. Jiao, S.; Chen, W.; Wei, G. Resilience and assemblage of soil microbiome in response to chemical contamination combined with plant growth. *Appl. Environ. Microbiol.* **2019**, *85*, e02523-18. [CrossRef]
14. Wang, Z.; Deng, G.; Hu, C.; Hou, X.; Zhang, X.; Fan, Z.; Zhao, Y.; Peng, M. Microbial diversity and community assembly in heavy metal-contaminated soils: Insights from selenium-impacted mining areas. *Front. Microbiol.* **2025**, *16*, 1561678. [CrossRef]
15. Pan, X.; Zhang, S.; Zhong, Q.; Gong, G.; Wang, G.; Guo, X.; Xu, X. Effects of soil chemical properties and fractions of Pb, Cd, and Zn on bacterial and fungal communities. *Sci. Total Environ.* **2020**, *715*, 136904. [CrossRef]
16. Liu, N.; Huang, X.; Sun, L.; Li, S.; Chen, Y.; Cao, X.; Wang, W.; Dai, J.; Rinnan, R. Screening stably low cadmium and moderately high micronutrients wheat cultivars under three different agricultural environments of China. *Chemosphere* **2020**, *241*, 125065. [CrossRef] [PubMed]
17. Li, J.; Zheng, Q.; Liu, J.; Pei, S.; Yang, Z.; Chen, R.; Ma, L.; Niu, J.; Tian, T. Bacterial-fungal interactions and response to heavy metal contamination of soil in agricultural areas. *Front. Microbiol.* **2024**, *15*, 1395154. [CrossRef] [PubMed]
18. Dusengemungu, L.; Gwanama, C.; Simuchimba, G.; Mubemba, B. Potential of bioaugmentation of heavy metal-contaminated soils in the Zambian Copperbelt using autochthonous filamentous fungi. *Front. Microbiol.* **2022**, *13*, 1045671. [CrossRef]
19. Lin, Y.; Ye, Y.; Hu, Y.; Shi, H. The variation in microbial community structure under different heavy metal contamination levels in paddy soils. *Ecotoxicol. Environ. Saf.* **2019**, *180*, 557–564. [CrossRef]
20. Chun, S.J.; Kim, Y.J.; Cui, Y.; Nam, K.H. Ecological network analysis reveals distinctive microbial modules associated with heavy metal contamination of abandoned mine soils in Korea. *Environ. Pollut.* **2021**, *289*, 117851. [CrossRef]
21. Zeng, X.Y.; Li, S.W.; Leng, Y.; Kang, X.H. Structural and functional responses of bacterial and fungal communities to multiple heavy metal exposure in arid loess. *Sci. Total Environ.* **2020**, *723*, 138081. [CrossRef]
22. Comte, J.; Fauteux, L.; del Giorgio, P.A. Links between metabolic plasticity and functional redundancy in freshwater bacterioplankton communities. *Front. Microbiol.* **2013**, *4*, 112. [CrossRef]
23. Silva, G.O.A.; Southam, G.; Gagen, E.J. Accelerating soil aggregate formation: A review on microbial processes as the critical step in a post-mining rehabilitation context. *Soil Res.* **2023**, *61*, 209–223. [CrossRef]
24. Chen, Y.; Tao, S.; Ma, J.; Qu, Y.; Sun, Y.; Wang, M.; Cai, Y. New insights into assembly processes and driving factors of urban soil microbial community under environmental stress in Beijing. *Sci. Total Environ.* **2024**, *947*, 174551. [CrossRef]
25. Sun, T.; Li, G.; Mazarji, M.; Delaplace, P.; Yang, X.; Zhang, J.; Pan, J. Heavy metals drive microbial community assembly process in farmland with long-termbiosolids application. *J. Hazard. Mater.* **2024**, *468*, 133845. [CrossRef]
26. Zhang, M.; Zhang, T.; Zhou, L.; Lou, W.; Zeng, W.; Liu, T.; Yin, H.; Liu, H.; Liu, X.; Mathivanan, K.; et al. Soil microbial community assembly model in response to heavy metal pollution. *Environ. Res.* **2022**, *213*, 113576. [CrossRef] [PubMed]
27. Yotova, G.; Padareva, M.; Hristova, M.; Astel, A.; Georgieva, M.; Dinev, N.; Tsakovski, S. Establishment of geochemical background and threshold values for eight potential toxic elements in the Bulgarian soil quality monitoring network. *Sci. Total Environ.* **2018**, *643*, 1297–1303. [CrossRef] [PubMed]
28. Giri, S.; Singh, A.K. Ecological and human health risk assessment of agricultural soils based on heavy metals in mining areas of Singhbhum copper belt, India. *Hum. Ecol. Risk Assess. Int. J.* **2017**, *23*, 1008–1027. [CrossRef]
29. Wen, C.; Wu, L.; Qin, Y.; Van Nostrand, J.D.; Ning, D.; Sun, B.; Xue, K.; Liu, F.; Deng, Y.; Liang, Y.; et al. Evaluation of the reproducibility of amplicon sequencing with Illumina MiSeq platform. *PLoS ONE* **2017**, *12*, e0176716. [CrossRef]
30. Bosch, J.; Némethová, E.; Tláskal, V.; Brabcová, V.; Baldrian, P. Bacterial, but not fungal, communities show spatial heterogeneity in European beech (*Fagus sylvatica* L.) deadwood. *FEMS Microbiol. Ecol.* **2023**, *99*, fiad023. [CrossRef]

31. Bolyen, E.; Rideout, J.R.; Dillon, M.R.; Bokulich, N.A.; Abnet, C.C.; Al-Ghalith, G.A.; Alexander, H.; Alm, E.J.; Arumugam, M.; Asnicar, F.; et al. Reproducible, interactive, scalable and extensible microbiome data science using QIIME 2. *Nat. Biotechnol.* **2019**, *37*, 852–857. [CrossRef]
32. Callahan, B.; McMurdie, P.; Rosen, M.; Han, A.; Johnson, A.; Holmes, A. DADA2: High-resolution sample inference from Illumina amplicon data. *Nat. Methods* **2016**, *13*, 581–583. [CrossRef] [PubMed]
33. Douglas, G.M.; Maffei, V.J.; Zaneveld, J.R.; Yurgel, S.; Brown, J.; Taylor, C.; Huttenhower, C.; Langille, M. PICRUSt2 for prediction of metagenome functions. *Nat. Biotechnol.* **2020**, *38*, 685–688. [CrossRef]
34. McDonald, D.; Price, M.N.; Goodrich, J.; Nawrocki, E.P.; DeSantis, T.Z.; Probst, A.; Andersen, G.L.; Knight, R.; Hugenholtz, P. An improved Greengenes taxonomy with explicit ranks for ecological and evolutionary analyses of bacteria and archaea. *ISME J.* **2012**, *6*, 610–618. [CrossRef] [PubMed]
35. McDonald, D.; Jiang, Y.; Balaban, M.; Cantrell, K.; Zhu, Q.; Gonzalez, A.; Morton, J.T.; Nicolaou, G.; Parks, D.H.; Karst, S.M.; et al. Greengenes2 unifies microbial data in a single reference tree. *Nat. Biotechnol.* **2024**, *42*, 715–718. [CrossRef]
36. Robeson, M.S., 2nd; O'Rourke, D.R.; Kaehler, B.D.; Ziemski, M.; Dillon, M.R.; Foster, J.T.; Bokulich, N.A. RESCRIPt: Reproducible sequence taxonomy reference database management. *PLoS Comput. Biol.* **2021**, *17*, e1009581. [CrossRef]
37. Nilsson, R.H.; Larsson, K.H.; Taylor, A.F.S.; Bengtsson-Palme, J.; Jeppesen, T.S.; Shigel, D.; Kennedy, P.; Picard, K.; Glöckner, F.O.; Tedersoo, L.; et al. The UNITE database for molecular identification of fungi: Handling dark taxa and parallel taxonomic classifications. *Nucleic Acids Res.* **2019**, *47*, D259–D264. [CrossRef] [PubMed]
38. Edgar, R.C.; Haas, B.J.; Clemente, J.C.; Quince, C.; Knight, R. UCHIME improves sensitivity and speed of chimera detection. *Bioinformatics* **2011**, *27*, 2194–2200. [CrossRef]
39. Hammer, Ø.; Harper, D.A.T.; Ryan, P.D. PAST: Paleontological statistics software package for education and data analysis. *Palaeontol. Electron.* **2001**, *4*, 9. Available online: http://palaeo-electronica.org/2001_1/past/issue1_01.htm (accessed on 13 May 2001).
40. Kot, A.; Namieśnik, J. The role of speciation in analytical chemistry. *Trends Anal. Chem.* **2000**, *19*, 69–79. [CrossRef]
41. Roane, T.M.; Josephson, K.L.; Pepper, I.L. Dual-bioaugmentation strategy to enhance remediation of cocontaminated soil. *Appl. Environ. Microbiol.* **2001**, *67*, 3208–3215. [CrossRef]
42. Knotek-Smith, H.M.; Deobald, L.A.; Ederer, M.; Crawford, D.L. Cadmium stress studies: Media development, enrichment, consortia analysis, and environmental relevance. *Biometals* **2003**, *16*, 251–261. [CrossRef]
43. Håkanson, L. An ecological risk index for aquatic pollution control—A sedimentological approach. *Water Res.* **1980**, *14*, 975–1001. [CrossRef]
44. Welp, G. Inhibitory effects of the total and water-soluble concentrations of nine different metals on the dehydrogenase activity of a loess soil. *Biol. Fertil. Soils* **1999**, *30*, 132–139. [CrossRef]
45. Sjöstedt, C.; Löv, Å.; Olivecrona, Z.; Boye, K.; Kleja, D.B. Improved geochemical modeling of lead solubility in contaminated soils by considering colloidal fractions and solid phase EXAFS speciation. *Appl. Geochem.* **2018**, *92*, 110–120. [CrossRef]
46. Torsvik, V.; Øvreås, L. Microbial diversity and function in soil: From genes to ecosystems. *Curr. Opin. Microbiol.* **2002**, *5*, 240–245. [CrossRef] [PubMed]
47. Rajapaksha, R.M.C.P.; Tobor-Kapłon, M.A.; Bååth, E. Metal toxicity affects fungal and bacterial activities in soil differently. *Appl. Environ. Microbiol.* **2004**, *70*, 2966–2973. [CrossRef]
48. Tipayno, S.C.; Truu, J.; Samaddar, S.; Truu, M.; Preem, J.K.; Oopkaup, K.; Espenberg, M.; Chatterjee, P.; Kang, Y.; Kim, K.; et al. The bacterial community structure and functional profile in the heavy metal contaminated paddy soils, surrounding a nonferrous smelter in South Korea. *Ecol. Evol.* **2018**, *8*, 6157–6168. [CrossRef]
49. Fajardo, C.; Costa, G.; Nande, M.; Botías, P.; García-Cantalejo, J.; Martín, M. Pb, Cd, and Zn soil contamination: Monitoring functional and structural impacts on the microbiome. *Appl. Soil Ecol.* **2019**, *135*, 56–64. [CrossRef]
50. Song, J.; Shen, Q.; Wang, L.; Qui, G.; Shi, J.; Xu, J.; Brookes, P.h.C.; Liu, X. Effects of Cd, Cu, Zn and their combined action on microbial biomass and bacterial community structure. *Environ. Pollut.* **2018**, *243*, 510–518. [CrossRef]
51. Schneider, A.R.; Gommeaux, M.; Duclercq, J.; Fanin, N.; Conreux, A.; Alahmad, A.; Lacoux, J.; Roger, D.; Spicher, F.; Ponthieu, M.; et al. Response of bacterial communities to Pb smelter pollution in contrasting soils. *Sci. Total Environ.* **2017**, *605–606*, 436–444. [CrossRef]
52. Kim, H.S.; Lee, S.H.; Jo, H.Y.; Finneran, K.T.; Kwon, M.J. Diversity and composition of soil Acidobacteria and Proteobacteria communities as a bacterial indicator of past land-use change from forest to farmland. *Sci. Total Environ.* **2021**, *797*, 148944. [CrossRef]
53. Mhete, M.; Eze, P.N.; Rahube, T.O.; Akinyemi, F.O. Soil properties influence bacterial abundance and diversity under different land-use regimes in semi-arid environments. *Sci. Afr.* **2020**, *7*, e00246. [CrossRef]
54. Wang, X.; Gao, P.; Li, D.; Liu, J.; Yang, N.; Gu, W.; He, X.; Tang, W. Risk assessment for and microbial community changes in farmland soil contaminated with heavy metals and metalloids. *Ecotoxicol. Environ. Saf.* **2019**, *185*, 109685. [CrossRef]

55. Beattie, R.E.; Henke, W.; Campa, M.F.; Hazen, T.C.; McAliley, L.R.; Campbell, J.H. Variation in microbial community structure correlates with heavy-metal contamination in soils decades after mining ceased. *Soil Biol. Biochem.* **2018**, *126*, 57–63. [CrossRef]
56. Prakash, A.A.; Rajasekar, A.; Sarankumar, R.K.; AlSalhi, M.S.; Devanesan, S.; Aljaafreh, M.J.; Govarthanan, M.; Saye, S.R.M. Metagenomic analysis of microbial community and its role in bioelectrokinetic remediation of tannery contaminated soil. *J. Hazard. Mater.* **2021**, *412*, 125133. [CrossRef]
57. Egidi, E.; Delgado-Baquerizo, M.; Plett, J.M.; Wang, J.; Eldridge, D.J.; Bardgett, R.D.; Maestre, F.D.; Singh, B.K. A few Ascomycota taxa dominate soil fungal communities worldwide. *Nat. Commun.* **2019**, *10*, 2369. [CrossRef] [PubMed]
58. Passarini, M.R.Z.; Robayo, M.I.G.; Ottoni, J.R.; Duarte, A.W.F.D.; Rosa, L.H. Biotechnological potential in agriculture of soil Antarctic microorganisms revealed by omics approach. *World J. Microbiol. Biotechnol.* **2024**, *40*, 345. [CrossRef]
59. Tedersoo, L.; Sánchez-Ramírez, S.; Kõljalg, U.; Bahram, M.; Döring, M.; Schigel, D.; May, T.; Ryberg, M.; Abarenkov, K. High-level classification of the Fungi and a tool for evolutionary ecological analyses. *Fungal Divers.* **2018**, *90*, 135–159. [CrossRef]
60. Yu, X.; Zhao, J.; Liu, X.; Sun, L.; Tian, J.; Wu, N. Cadmium Pollution Impact on the Bacterial Community Structure of Arable Soil and the Isolation of the Cadmium Resistant Bacteria. *Front. Microbiol.* **2021**, *12*, 698834. [CrossRef]
61. Zou, L.; Lu, Y.; Dai, Y.; Khan, M.I.; Gustave, W.; Nie, J.; Liao, Y.; Tang, X.; Shi, J.; Xu, J. Spatial variation in microbial community in response to As and Pb contamination in paddy soils near a Pb–Zn mining site. *Front. Environ. Sci.* **2021**, *9*, 630668. [CrossRef]
62. Feng, G.; Xie, T.; Wang, X.; Bai, J.; Zhao, H.; Wei, W.; Wang, M.; Zhao, Y. Metagenomic analysis of microbial community and function involved in Cd-contaminated soil. *BMC Microbiol.* **2018**, *18*, 11. [CrossRef]
63. Maheshwari, P.; Murali Sankar, P. Culture-independent and culture-dependent approaches in symbiont analysis: In proteobacteria. In *Microbial Symbionts*; Developments in Applied Microbiology and Biotechnology; Elsevier: Amsterdam, The Netherlands, 2023; Chapter 42; pp. 743–763. [CrossRef]
64. Saxena, A.K.; Kumar, M.; Chakdar, H.; Anuroopa, N.; Bagyaraj, D.J. Bacillus species in soil as a natural resource for plant health and nutrition. *J. Appl. Microbiol.* **2019**, *128*, 1583–1594. [CrossRef]
65. Wimalasekara, R.L.; Seneviratne, K.N.; Jayathilaka, N. Metagenomics in bioremediation of metals for environmental cleanup. In *Metagenomics to Bioremediation*; Developments in Applied Microbiology and Biotechnology; Elsevier: Amsterdam, The Netherlands, 2023; Chapter 9; pp. 231–259. [CrossRef]
66. Guo, H.; Nasir, M.; Lv, J.; Dai, Y.; Gao, J. Understanding the variation of microbial community in heavy metals contaminated soil using high throughput sequencing. *Ecotoxicol. Environ. Saf.* **2017**, *144*, 300–306. [CrossRef]
67. Guo, Y.; Cheng, S.; Fang, H.; Yang, Y.; Li, Y.; Zhou, Y. Responses of soil fungal taxonomic attributes and enzyme activities to copper and cadmium co-contamination in paddy soils. *Sci. Total Environ.* **2022**, *844*, 157119. [CrossRef] [PubMed]
68. Ye, F.; Gong, D.; Pang, C.; Luo, J.; Zeng, X.; Shang, C. Analysis of fungal composition in mine-contaminated soils in Hechi City. *Curr. Microbiol.* **2020**, *77*, 2685–2693. [CrossRef] [PubMed]
69. Mohammadian, E.; Babai Ahari, A.; Arzanlou, M.; Oustan, S.; Khazaei, S.H. Tolerance to heavy metals in filamentous fungi isolated from contaminated mining soils in the Zanjan Province, Iran. *Chemosphere* **2017**, *185*, 290–296. [CrossRef]
70. Montes-Montes, G.; Munoz-Ramirez, Z.Y.; Cortes-Palacios, L.; Carrillo-Campos, J.; Ramirez-Sanchez, O.; Ortiz-Aguirre, I.; Munoz-Castellanos, L.N.; Gonzalez-Escobedo, R. Microbial Diversity and Heavy Metal Resistome in Slag-Contaminated Soils from an Abandoned Smelter in Chihuahua, Mexico. *Soil Syst.* **2025**, *9*, 30. [CrossRef]
71. Lemmel, F.; Maunoury-Danger, F.; Leyval, C.; Cébron, A. Altered fungal communities in contaminated soils from French industrial brownfields. *J. Hazard. Mater.* **2021**, *406*, 124296. [CrossRef]
72. Chen, J.; Li, J.; Zhang, H.; Shi, W.; Liu, Y. Bacterial heavy-metal and antibiotic resistance genes in a copper tailing dam area in northern China. *Front. Microbiol.* **2019**, *10*, 1916. [CrossRef] [PubMed]
73. Goswami, A.; Adkins-Jablonsky, S.J.; Barreto Filho, M.M.; Shilling, M.D.; Dawson, A.; Heiser, S.; O'Connor, A.; Walker, M.; Roberts, Q.; Morris, J.J. Heavy metal pollution impacts soil bacterial community structure and antimicrobial resistance at the Birmingham 35th Avenue Superfund site. *Microbiol. Spectr.* **2023**, *11*, e02426-22. [CrossRef] [PubMed]
74. Nies, D.H. Efflux-mediated heavy metal resistance in prokaryotes. *FEMS Microbiol. Rev.* **2003**, *27*, 313–339. [CrossRef] [PubMed]
75. Luo, L.; Xie, L.; Jin, D.; Mi, B.; Wang, D.; Li, X.; Dai, X.; Zou, X.; Zhang, Z.; Ma, Y.; et al. Bacterial community response to cadmium contamination of agricultural paddy soil. *Appl. Soil Ecol.* **2019**, *139*, 100–106. [CrossRef]

toxics

Article

Concentration-Dependent Effects of Polyethylene Microplastics on Cadmium and Lead Bioavailability in Soil

Zhenbo Wang [1,2], Sihan Liu [1,2], Peng Zhao [1,2,*], Guangxin Li [1,2], Ran Duan [3], Chang Li [1,2] and Haichao Fu [1,2,*]

1 College of Resources and Environmental Sciences, Henan Agricultural University, Zhengzhou 450002, China
2 Key Laboratory of Soil Pollution Control and Remediation of Henan Province, Zhengzhou 450002, China
3 Institute of Quality and Safety for Agro-Products, Henan Academy of Agricultural Sciences, Zhengzhou 450002, China
* Correspondence: zhpddy@163.com (P.Z.); haichaofu@henau.edu.cn (H.F.)

Abstract

The influence of microplastics (MPs) on the availability of soil heavy metals (HMs) is a current research hotspot, but how MPs regulate HM availability via soil properties and the bacterial community remains unclear. This study investigated the effects of polyethylene (PE) MP concentrations on soil properties, bacterial communities, surface chemistry, and the speciation of cadmium (Cd) and lead (Pb) through soil incubation. Results indicated that as PE MP concentration increased, soil pH and cation exchange capacity declined, while organic carbon concentration increased. Available phosphorus and alkali–hydrolyzable nitrogen concentrations increased at 0.1% and 1% PE MPs, but decreased at 10% PE MPs. Bacterial community indices, including Simpson, ACE, and Chao1, increased at 0.1% and 1% PE MPs but decreased at 10% PE MPs. PE MPs (0.1% and 1%) reduced DTPA–Cd/Pb, promoting their transformation into stable forms and surface complexation with oxygen–containing groups. In contrast, 10% PE MPs disrupted the formation of PbO, $PbCO_3$, and $Cd(OH)_2$, producing the opposite effect. The random forest model revealed that soil organic carbon and available phosphorus were the primary factors influencing DTPA–Pb and DTPA–Cd, respectively. Partial least squares path modeling demonstrated that PE MPs altered the physicochemical characteristics of soil and structure of bacterial communities, ultimately impacting transformation of Cd and Pb speciation, with these changes being highly dependent on PE MP concentration.

Keywords: cadmium; lead; polyethylene microplastics; soil properties; bacterial community structure; availability

1. Introduction

The advancement of contemporary industry has resulted in a substantial discharge of heavy metals (HMs) into the soil, leading to serious pollution of agricultural ecosystems [1]. HMs in cropland can be readily absorbed by crops and enter the human body through food intake, posing a significant risk to human health [2–4]. Among the various HMs, lead (Pb) and cadmium (Cd) are particularly concerning due to their chronic and acute toxicological effects on animals and plants [5,6]. Soil serves as a primary reservoir for various pollutants, including HMs and other contaminants. Polyethylene (PE) is the most widely used plastic polymer globally, demonstrating significant application value in greenhouse construction and plastic mulching [7]. Due to its difficulty in being fully recycled and reused, significant accumulation of these plastics has occurred in the environment. Plastic waste undergoes a range of physical, chemical, and microbiological processes that generate particles smaller than 5 mm, collectively referred to as microplastics (MPs), among which polyethylene microplastics (PE MPs) are a major type [8,9]. MPs are widely distributed across various land use types globally, with abundance ranges of 0–10^6 items/kg and 0–10^5 mg/kg. Agricultural soils exhibit a higher prevalence of MP contamination (61%) [10]. Liu et al. [11] reported that the abundance of MPs in agricultural soils has reached levels ranging from 4350 items/kg to 24,120 items/kg. Notably, low–density polyethylene (LDPE) plastic, with a size of 60 μm, requires 300 years to fully degrade in soil [7], indicating that MPs in soil gradually accumulate, leading to long–term pollution. Furthermore, the rapid increase in plastic production and mineral resource exploitation has resulted in the spatial overlap of MP and HM pollution [12]. Therefore, investigating the relationship between PE MPs and the availability of soil HMs is of paramount importance.

Soil physicochemical properties, including pH, cation exchange capacity (CEC), and soil organic carbon (SOC), are key regulators of heavy metal behavior in soils. pH influences the adsorption sites, coordination characteristics, chemical structure, and surface stability of HMs in soil [13], thereby altering their morphological properties and availability [14]. Furthermore, a decrease in CEC leads to increased availability of HMs in the soil [13]. SOC, rich in ligands and functional groups, plays an essential role in forming complexes with metal ions, which in turn influences the transport and transformation of HM forms in the soil environment [15,16]. Previous research has demonstrated that MPs alter the effectiveness of HMs in soil ecosystems, with the degree and direction of these effects contingent upon the concentration levels of MPs. For instance, Wang et al. [17] demonstrated that 10% polyethylene supplementation significantly decreased DTPA–extractable Cd in soil compared to 0.1% and 1% PE MPs supplementation. Conversely, another study indicated that 10% PE MPs resulted in elevated available Cd levels in soil compared to lower doses [14]. The findings from these studies are inconsistent; thus, the joint impacts of MPs and HMs require further examination, particularly concerning the possible concentration–effect correlation.

The presence of HMs and MPs in soil can exert various direct or indirect effects on soil microorganisms [18]. Research indicated that simultaneous exposure to plastic microfibers (PMFs) and Cd leads to an increased relative abundance of the bacterial genus *Sphingomonas*, while concurrently suppressing iron–oxidizing bacteria [19]. A study conducted by Tang et al. [20], polyethylene terephthalate (PET) MPs decrease the accessibility of HMs but also change the makeup of soil microbial communities in soils contaminated with HMs. Furthermore, findings by Jiang et al. [21] revealed that the presence of both MPs and Cd in the soil environment disrupts the structure of bacterial communities. Additional evidence from Feng et al. [22] suggested that traditional MPs and biodegradable MPs not only restructure microbial communities in soils contaminated with HMs but also interfere with functional genes associated with exogenous biodegradation and metabolism, consequently affecting the availability of lead (Pb) and zinc (Zn). Previous research has primarily focused on the external addition of HMs to soil [19,22,23], but how MPs affect bacterial communities in situ Cd–Pb contaminated soils has rarely been investigated.

Previous research has produced contradictory conclusions on how MPs affect the bioavailability of HMs in soils, with some studies reporting decreases [17], others increases [14], and some no significant effects [19]. Moreover, most existing work has overlooked the potential concentration–dependent effects of MPs and rarely integrated soil physicochemical properties, microbial communities, and molecular–level mechanisms into a single framework. To address this gap, the present study (i) reveals a concentration–dependent regulation of Cd and Pb bioavailability by polyethylene MPs, showing opposite effects at low versus high doses of PE MPs; (ii) employs a combination of soil chemical indices, bacterial community analysis, and advanced spectroscopic techniques (Fourier transform infrared spectroscopy, X–ray photoelectron spectroscopy, X–ray diffraction) to elucidate the underlying mechanisms; and (iii) applies random forest and partial least squares path modeling to disentangle the relative contributions of soil properties and microbial communities to HM availability. This integrative approach fills an important research niche by clarifying how varying concentrations of polyethylene MPs regulate HM speciation and availability in co–contaminated agricultural soils.

2. Material and Methods

2.1. PE MPs and Soils

The PE MPs were characterized before the incubation experiment (Figure S1). Their morphology was examined using scanning electron microscopy (SEM, ZEISS Sigma 300, Oberkochen, Germany) and confirmed to consist of irregular fragments measuring between 5 and 15 μm (Figure S1A). Fourier transform infrared spectroscopy (FTIR, Nicolet iS10, Thermo Fisher Scientific, Waltham, MA, USA) was employed to verify the polymer composition by matching the characteristic absorption bands of polyethylene at approximately 2915, 2864, 1470, and 717 cm^{-1} (Figure S1B) [24].

The Cd– and Pb–contaminated soil used in this study was collected from agricultural land in Jiyuan City, Henan Province, China. The samples were taken from the topsoil (0–20 cm), and the tested soil type was cinnamon soil. The concentrations of available Cd and Pb in the soil were 1.22 mg/kg and 56.39 mg/kg, respectively. The soil was allowed to dry in the air at room temperature before being sifted through a 2 mm mesh for further use. Table S1 displays the physicochemical properties of the soil sample.

2.2. Soil Incubation Experiment

Soil incubation experiments were conducted with four treatments: 0% PE MPs (control, CK), 0.1% PE MPs (0.1%), 1% PE MPs (1%), and 10% PE MPs (10%), each with twelve replicates, requiring 200 g of dry soil per repetition. Specifically, MPs were thoroughly ground and mixed with a small quantity of soil, which was then gradually incorporated into the remaining 180 g of soil. This mixture was stirred repeatedly with a glass rod for 20 min to ensure even distribution of the MPs throughout the soil. During the incubation period, deionized water was added regularly to maintain 60% of the optimal water–holding capacity. Soil samples were collected after 10, 30, 60, and 90 days, with three replicates randomly collected from each sampling time. At 90 days, fresh soil samples were obtained for bacterial diversity analysis, while the leftover samples were air–dried, sifted through 20–mesh and 100–mesh sieves, and subsequently stored in bags for later use.

2.3. Soil Physicochemical Analysis

Soil pH was measured using a pH meter with a suspension ratio of 1:2.5 (soil: water, *w*/*v*) [25]. The diffusion method was employed to determine alkali–hydrolyzable nitrogen (AH–N) [26]. To extract soil available phosphorus (AP), sodium bicarbonate was utilized, and its concentration was then measured using the molybdenum–antimony colorimetric technique [26]. For evaluating the soil cation exchange capacity, the spectrophotometric method involving hexamminecobalt trichloride solution was applied (HJ 889−2017). The quantification of SOC was conducted through the potassium dichromate–sulfuric acid oxidation method. The available Cd and Pb were obtained using a diethylene triamine penta–acetic acid (DTPA) solution at a concentration of 0.005 mol/L, with a pH of 7.3 [25], and were measured with atomic absorption spectrophotometer (ZEEnit700, Analytik Jena AG, Jena, Germany).

The approach developed by Tessier et al. [27] was employed to investigate the various forms of Cd and Pb found in the soil. Additional information regarding the analysis of Cd speciation in soil is available in Supplementary Materials Text S1.

2.4. Soil DNA Extraction and 16S rRNA Illumina Sequencing

Genomic DNA extraction was performed on microbial sources by utilizing soil samples processed with the E.Z.N.A® Soil DNA Kit, providing the template for amplifying the hypervariable V3–V4 region of the bacterial 16S rRNA gene with primer pairs 338F and 806R. Once the amplification was complete, the PCR product was retrieved from a 2% agarose gel and purified, after which its concentration was measured using a Qubit 4.0 (Thermo Fisher Scientific, Waltham, MA, USA). The purified amplicons were combined in equal molar concentrations and subjected to paired–end sequencing on the Illumina Nextseq2000 platform (Illumina, San Diego, CA, USA), following the standard protocols established by Majorbio Bio–Pharm Technology Co. Ltd. (Shanghai, China). To maintain sample integrity, quality control of the paired–end raw sequencing reads was conducted using the fastp software (https://github.com/OpenGene/fastp (accessed on 20 July 2024, version 0.19.6). Subsequently, the processed sequences were clustered into operational taxonomic units (OTUs) with UPARSE 7.1 (http://drive5.com/uparse/ (accessed on 20 July 2024)) at a 97% sequence similarity threshold, with the most prevalent sequence in each OTU chosen as the representative sequence.

2.5. Characterization of Soil Components

The crystallographic structures of the soil were examined following a 90–day incubation period using X–ray diffraction (XRD, D8 Advance, Bruker, Ettlingen, Germany), which conducted at a scanning speed of 2°/min within the range of $10° \leq 2\theta \leq 80°$. Additionally, the crystallographic characteristics of the soil particles were evaluated through X–ray photoelectron spectroscopy (XPS, Thermo Scientific K–Alpha, Waltham, MA, USA). For Fourier transform infrared spectroscopy (FTIR), soil samples were air–dried and ground using an agate mortar to pass through a 100–mesh sieve. Approximately 1 mg of the finely ground soil sample was thoroughly mixed with 100 mg of spectroscopic grade KBr, ground again, and pressed into a thin sheet under vacuum. The pellets were scanned in the range of 4000–400 cm^{-1} with a resolution of 4 cm^{-1} using a Thermo Scientific Nicolet iS10 FTIR spectrometer (FTIR, Nicolet iS10, Thermo Fisher Scientific, Waltham, MA, USA) [28]. These techniques (FTIR, XRD, and XPS) were employed to identify specific changes in functional groups, mineral phases, and surface elemental composition relevant to soil–MPs–HMs interactions, as widely reported in previous studies [23,29].

2.6. Statistical Analysis

A one–way analysis of variance (ANOVA) was performed with the aid of IBM SPSS Statistics 26.0 software (IBM SPSS Inc., Chicago, USA), where means were evaluated using Duncan's multiple range tests. Line graphs and circular bar plots were generated employing Origin software 2024 (OriginLab Corp., Northampton, MA, USA). Bioinformatic analysis of soil microbiota and the Mantel test were conducted through the Majorbio Cloud platform (https://cloud.majorbio.com (accessed on 25 July 2024)). The Spearman correlation analysis was carried out utilizing Wekemo Bioincloud (https://www.bioincloud.tech(accessed on 25 July 2024)). Random forest models were created using R software (version 4.3.2; R Core Team) to identify the key microorganisms affecting changes in the soil environment. Furthermore, Partial least squares path modeling (PLS–PM) was applied to explore potential associations among microplastic concentration, soil chemical index, alpha diversity index, and the speciation and bioavailability of Cd/Pb, also with R software.

3. Results

3.1. Effect of PE MPs on Available Cd and Pb

The incorporation of PE MPs has a large effect on the availability of Cd and Pb in soil (Figure 1). During the initial stages of incubation, the introduction of PE MPs resulted in lower levels of DTPA–Cd and DTPA–Pb. However, as the incubation period progressed, the application of 10% PE MPs led to a significant increase in the levels of DTPA–Cd and DTPA–Pb. Following a 90–day incubation period, compared to CK, DTPA–Cd concentrations decreased by 12.21% and 10.14% under the 0.1% and 1.0% PE MP treatments, respectively. In contrast, DTPA–Cd concentrations increased by 63.74% under the 10% PE MPs treatment (Figure 1A). The effect of PE MPs on DTPA–Pb mirrored that on DTPA–Cd. Following a 90–day incubation period, compared to CK, DTPA–Pb concentrations decreased by 19.76% and 31.10% under the 0.1% and 1.0% PE MPs treatments, respectively, while DTPA–Pb concentrations increased by 44.69% under the 10% PE MPs treatment (Figure 1B).

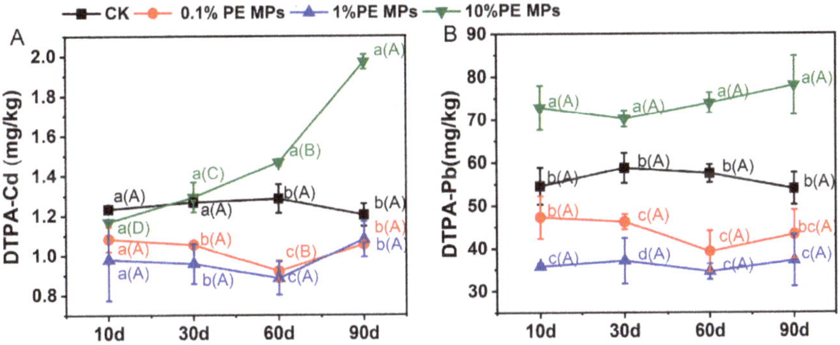

Figure 1. Effect of PE MPs on soil DTPA–Cd (**A**) and DTPA–Pb (**B**). Differences between various treatments at the same time points are indicated by lowercase letters, whereas uppercase letters denote differences across different time points. Data are presented as mean ± SD ($n = 3$). Error bars represent SD. The different letters denote significant differences ($p < 0.05$).

3.2. Effect of PE MPs on the Chemical Speciation of Cd and Pb

As shown in Figure 2, at 0.1% and 1% PE MPs, the exchangeable forms (F1) of Cd and Pb exhibited a gradual reduction over a 90–day incubation period. Specifically, the F1 percentage of Cd diminished from 26.35% to 10.46% and 9.51%, respectively. In contrast, the F1 percentage of Pb reduced from 15.21% to 9.02% and 6.11%, respectively, when compared to CK. At high concentration (10%), the F1 of Cd and Pb gradually increased, whereas after 90 days of incubation, the F1 of Cd and Pb increased from 26.35% to 36.35%, and from 15.21% to 24.78%, respectively, compared to CK. The carbonate–bound (F2) of Cd and Pb increased at 0.1% and 1% PE MPs but decreased at 10% PE MPs. The Fe–Mn oxide–bound (F3) of Cd and Pb decreased more significantly at 10% PE MPs. The (organic–bound) F4 of Cd and Pb was lowest in CK treatment, gradually increasing with high concentration of PE MPs and longer incubation times. After 90 days incubation phase, the F4 of Cd increased from 5.34% to 10.00%, 11.68%, and 15.13%, while the F4 of Pb increased from 11.21% to 17.65%, 19.53%, and 21.60%, respectively, compared to CK. The residual (F5) of Cd was lowest in the CK treatment and highest at 10% PE MPs, whereas the F5 of Pb was lower at 0.1%, 1% PE MPs and highest at 10% PE MPs.

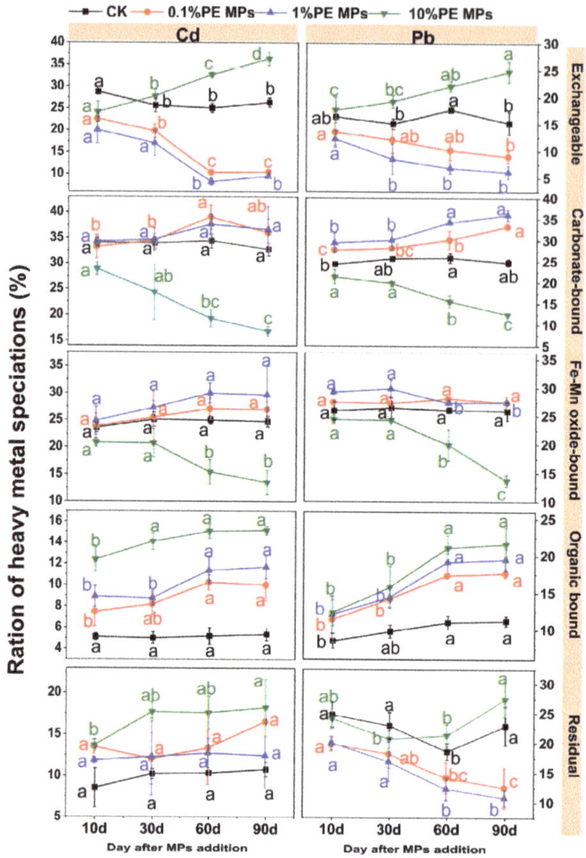

Figure 2. Effect of PE MPs on the chemical speciation of Cd and Pb. The data is expressed as mean ± SD (n = 3), with the error bars denoting the SD. The different letters denote significant differences ($p < 0.05$).

3.3. *Effects of PE MPs on Soil Physicochemical Properties*

The incorporation of PE MPs altered the characteristics of the soil. As illustrated in Figure 3A, soil pH consistently declined with increasing concentrations of PE MPs. After 90 days, the soil pH decreased by 6.62%, 8.24%, and 15.46% compared to the CK treatment, with statistical significance ($p < 0.05$). Under the 10% PE MPs treatment, soil pH steadily decreased over time. Specifically, compared to the pH at 10 days of incubation, the pH dropped by 0.42 units after 90 days. In contrast, the 0.1% and 1% PE MPs treatments caused minimal disturbance, and the soil pH remained stable throughout the incubation period (Figure S2). After 90 days, the 10% PE MPs treatment reduced CEC by 4.09% compared to CK. Furthermore, the SOC content significantly increased with rising PE MP concentrations (Figure 3C). After 90 days, SOC content increased by factors of 1.49, 2.42, and 9.20 compared to CK. Additionally, AH–N and AP content were also measured. As shown in Figure S3, AP content increased with 0.1% and 1% PE MPs, while the 10% PE MPs treatment had the opposite effect. Notably, the 1% PE MPs treatment significantly increased AH–N content compared to CK, whereas the 10% PE MPs treatment resulted in the lowest levels of AH–N.

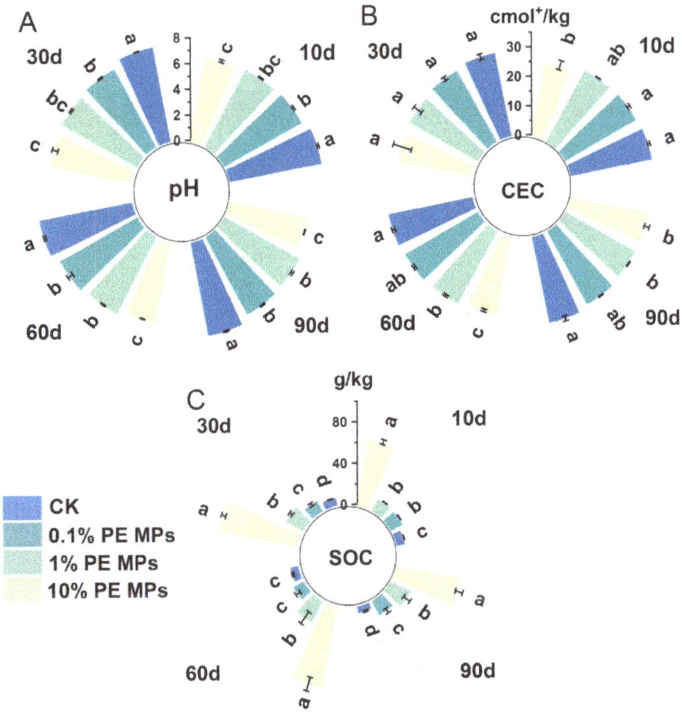

Figure 3. Effects of PE MPs on soil physicochemical properties ((**A**) pH, (**B**) CEC, (**C**) Soil organic carbon). The data is expressed as mean ± SD (n = 3), with the error bars denoting the SD. The different letters denote significant differences ($p < 0.05$).

3.4. Effects of PE MPs on the Diversity and Structure of Soil Bacterial Community

According to the results obtained from high–throughput sequencing of 16S rRNA, the quantity of OTUs at the genus level ranged from 728 in the 10% to 840 in the 0.1% PE MPs treatment (Figure 4A). The unique OTU counts for the various treatments were 32 (4.02%), 24 (2.99%), 34 (4.67%), and 40 (4.76%), with a total of 608 shared OTUs (19.19%) (Figure 4A and Figure S4). The addition of PE MPs influenced bacterial community diversity. The 1% PE MPs treatment increased the Simpson index, while the 10% PE MPs reduced both the Chao1 and ACE indices (Figure 4B). Principal coordinate analysis (PCoA) demonstrated that variations in the concentration of PE MPs, directly influenced the differences noted within the soil bacterial community (Figure 4C). At the phylum level, *Proteobacteria* and *Acidobacteriota* were the predominant phyla in the CK, 0.1%, and 1% PE MPs samples, whereas *Proteobacteria* and *Actinobacteriota* were the most prominent phyla in the 10% PE MPs sample. With 1% PE MPs, the relative abundance of *Cyanobacteria* increased, whereas under 10% PE MPs, the abundance of *Actinobacteriota* rose, and the abundance of *Acidobacteriota* declined (Figure S5B). At the genus level, bacterial communities exposed to the 0.1% PE MPs and CK treatments displayed similarities; however, the introduction of 1% PE MPs resulted in a decrease in the relative prevalence of *Bacillus* while simultaneously increasing the relative prevalence of *Cyanobacteriales* and *Leptolyngbya_ANT.L52.2*. After adding 10% PE MPs, there was an observed increase in the relative abundance of *Bacillus*, *Sphingomonas*, *Arthrobacter*, and *Saccharimonadales*, whereas the relative abundance of *RB41*, *Vicinamibacteraceae*, and *Vicinamibacterales* diminished (Figure 4D). According to the random forest model, *Elsterales*, *Marinococcaceae*, and *Sandaracinaceae* were identified as the three most important bacterial genera (Figure S5A). The abundance of *Brevundimonas* and *Obscuribacteraceae* increased at 0.1% PE MPs, while the abundance of *Methyloligellaceae* increased under 1% PE MPs.

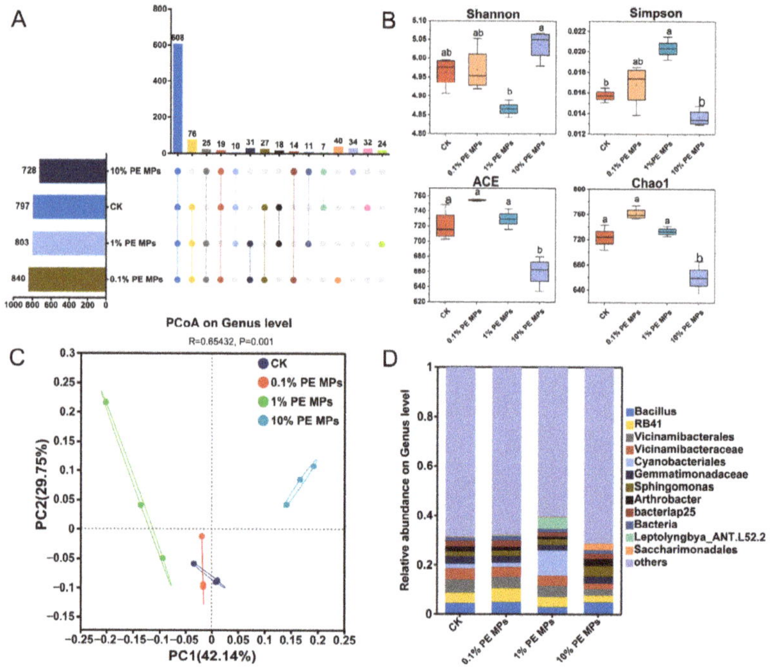

Figure 4. Effects of PE MPs on the diversity and structure of soil bacterial community. (**A**) Upset diagrams analysis of sediments treated with different PE MP concentrations. (**B**) Alpha diversity indices of the soil microbial community, the different letters denote significant differences ($p < 0.05$). (**C**) Principal Coordinates Analysis (PCoA) plot at the Genus level. (**D**) The relative abundance of bacterial community in soil at the genus level. Different colored nodes represent different phylum levels.

3.5. Correlation Analysis in Soil Physicochemical Properties and Key Bacterial

The analysis of bacterial communities at the phylum level identified four dominant phyla: *Proteobacteria*, *Acidobacteriota*, *Actinobacteriota*, and *Chloroflexi* in the soil (Figure 5). Their correlations with environmental factors were assessed using the Mantel test. For Cd, *Actinobacteriota* significantly influenced all factors except pH, CEC, Cd–F4, and Cd–F5 ($p < 0.01$). *Chloroflexi* significantly affected AP, AH–N, and Cd–F1 ($p < 0.05$), while *Proteobacteria* significantly impacted only Cd–F2 and SOC ($p < 0.05$) (Figure 5A). In the case of Pb, neither *Proteobacteria* nor *Actinobacteriota* had a significant effect on CEC and Pb–F4 ($p > 0.05$). Conversely, *Chloroflexi* significantly influenced Pb–F2 and Pb–F5 ($p < 0.05$), and *Acidobacteriota* significantly affected DTPA–Pb, SOC, Pb–F2, and Pb–F3 ($p < 0.05$) (Figure 5D). The redundancy analysis (RDA) indicated that soil properties accounted for 92.84% of the total variance in microbial communities for Cd, with the first and second axes (RDA1 and RDA2) explaining 76.15% and 16.69%, respectively. Specifically, Cd–F1, Cd–F2, Cd–F3, SOC, AH–N, and DTPA–Cd were the primary factors influencing microbial community distribution (Figure 5B). For Pb, the total variance was explained by RDA1 and RDA2 at rates of 70.13% and 14.74%, respectively, with the primary environmental variables influencing bacterial communities being Pb–F1, Pb–F2, Pb–F3, SOC, AH–N, and DTPA–Pb (Figure 5D).

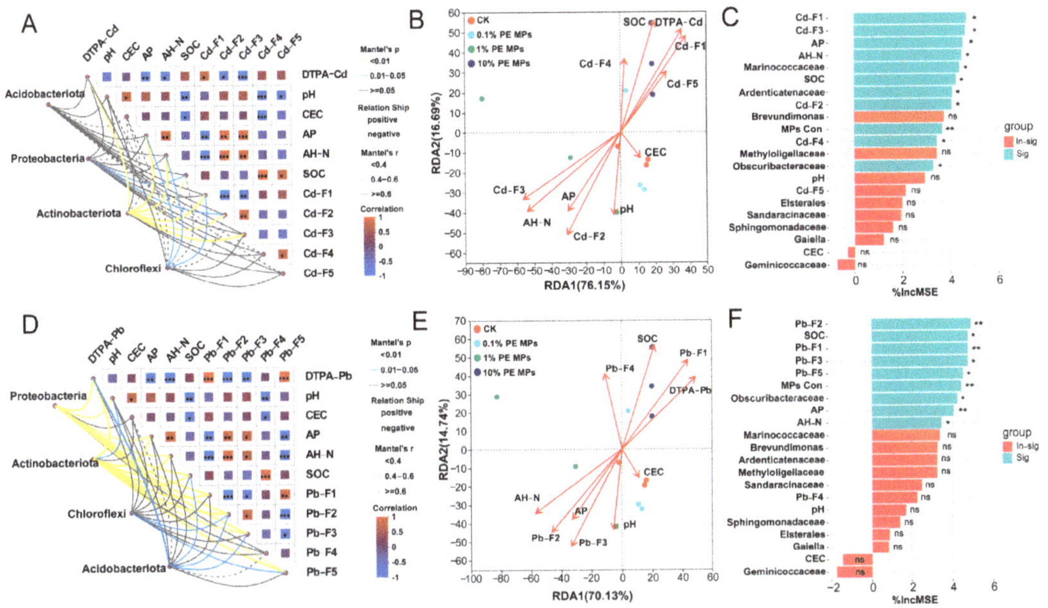

Figure 5. Analysis of the correlation between soil properties and bacterial community structure. Utilizing the Mantel test to examine the connection between soil physicochemical properties and the *Proteobacteria*, *Acidobacteriota*, *Actinobacteriota*, and *Chloroflexi* ((**A**) Cd, (**D**) Pb). Distribution pattern of soil physicochemical parameters by redundancy analysis (RDA) ((**B**) Cd, (**E**) Pb). Random forest analysis demonstrating the influence of soil properties and the ten most prevalent bacterial genera on DTPA–Cd (**C**) and DTPA–Pb (**F**). Significant levels are indicated: * $p < 0.05$; ** $p < 0.01$; *** $p < 0.001$.

The random forest model was employed to evaluate the influence of soil physicochemical properties and top 10 bacterial genera on DTPA–Cd and DTPA–Pb. The findings demonstrated that Cd–F1, Cd–F3, AP, and AH–N were the top 4 contributors to DTPA–Cd, while the genera *Methyloligellaceae*, *Ardenticatenaceae*, and *Obscuribacteraceae* had a notable impact on DTPA–Cd ($p < 0.05$). MPs concentration was identified as an important significant factor affecting bioavailable Cd content ($p < 0.01$) (Figure 5C). For DTPA–Pb, Pb–F2, SOC, Pb–F1, and Pb–F3 were the primary factors, with the genus *Obscuribacteraceae* contributing most significantly to bioavailable Pb content (Figure 5F). Pb–F2, Pb–F1, MPs–Con (MP concentration), and AP all significantly contributed to DTPA–Pb ($p < 0.01$). The top 10 bacterial genera were also strongly correlated with soil physicochemical properties. Significant positive relationships were observed between *Marinococcaceae*, *Methyloligellaceae* and AH–N, AP, whereas negative relationships were noted with the DTPA–Cd and DTPA–Pb ($p < 0.05$). *Sandaracinaceae* showed significant negative relationships with available nitrogen and phosphorus and significant positive relationships with bioavailable Cd and Pb ($p < 0.05$) (Figure S7).

3.6. Characterization of Soil

To further understand the effect of PE MPs in soil on the chemical form of Cd and Pb, FTIR, XPS and XRD data were collected from soil samples that were incubated for 90 days. FTIR analysis indicated an increase in the relative intensity of the –OH stretch at approximately 3226 cm^{-1} and the Si–O–Si band at around 1091 cm^{-1} under the influence of 10% PE MPs, while changes in the Fe–O band at approximately 470 cm^{-1} were also observed (Figure 6A) [29]. XRD indicated attenuation of reflections at $2\theta \approx 22.08°$ (PbO, PDF#76–2065) and 30.96° ($PbCO_3/Cd(OH)_2$, PDF#85–1088/PDF#71–2137) at 10% PE MPs

(Figure 6B). The atomic percentages of elements in CK, 0.1% PE MPs, 1% PE MPs, and 10% PE MPs were assessed through a comprehensive XPS analysis, as presented in Table S2. The atomic percentage of C1s increases with higher PE MP concentrations, while the atomic percentages of O1s and the O/C ratios decrease accordingly. The atomic percentages of Si 2p, Al 2p, and Fe 2p exhibited a trend that initially increased before subsequently decreasing as the PE MP concentration rose. As illustrated in Figure 7C, the peak intensity of O1s, Si 2p, and Fe 2p decreased slightly under 10% PE MPs treatment, whereas the intensity of the C1s peak increased slightly. These results suggest that the presence of PE MPs may induce physicochemical interactions at the soil–PE MPs interface, primarily involving hydrogen bonding, electrostatic attraction, and surface complexation with oxygen–containing functional groups (–OH, Si–O–Si, Fe–O), thereby altering the speciation of Cd and Pb in the soil.

Through XPS fine spectrum analysis, it was observed that after incubating soil samples for 90 days, there was an increase in the peaks of O=C–O in the C1s spectra, while the peak of C–C gradually decreased under 10% PE MPs (Figure 6D). The O1s spectrum indicated that at 0.1% and 1% PE MPs, the peaks of –OH significantly increased, with the –OH content on the surface of the soil fractions rising from 47.17% to 85.12% and 78.90%, respectively. In contrast, at 10% PE MPs, the peak of –OH decreased from 47.17% to 43.60%. Conversely, at 0.1% and 1% PE MPs, the C–O and O=C–O content on the soil surface decreased from 29.88% and 22.94% (CK) to 11.53% and 18.98% (0.1% PE MPs), and 2.35% and 5.12% (1% PE MPs), respectively. However, at 10% PE MPs, there was an increase in the peaks of C–O and O=C–O in the O1s spectra compared to 0.1% and 1% PE MPs. Notably, the binding energy of the O=C–O peak decreased from 288.7 eV to 287.5 eV in the C1s spectra with increasing MP concentrations after 90 days of incubation. These findings suggest that the impacts of different concentrations of PE MPs on soil functional groups vary significantly, with the key groups that are more strongly affected being primarily C–O, O=C–O, and –OH. This variability may be a crucial factor in the alterations of soil heavy metal chemical forms influenced by different concentrations of PE MPs.

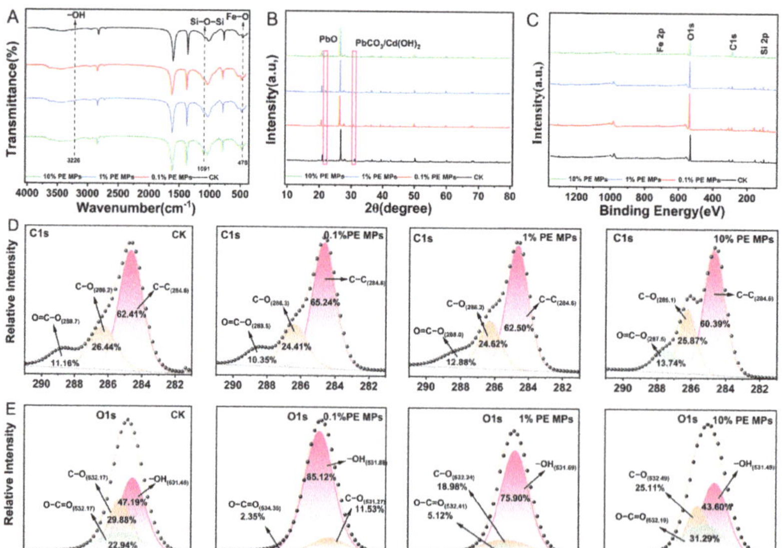

Figure 6. Characterization of after 90 days of incubation under different treatments. FTIR spectra (**A**), XRD patterns (**B**) and XPS spectra ((**C**) full survey; (**D**) C 1 s; (**E**) O 1 s).

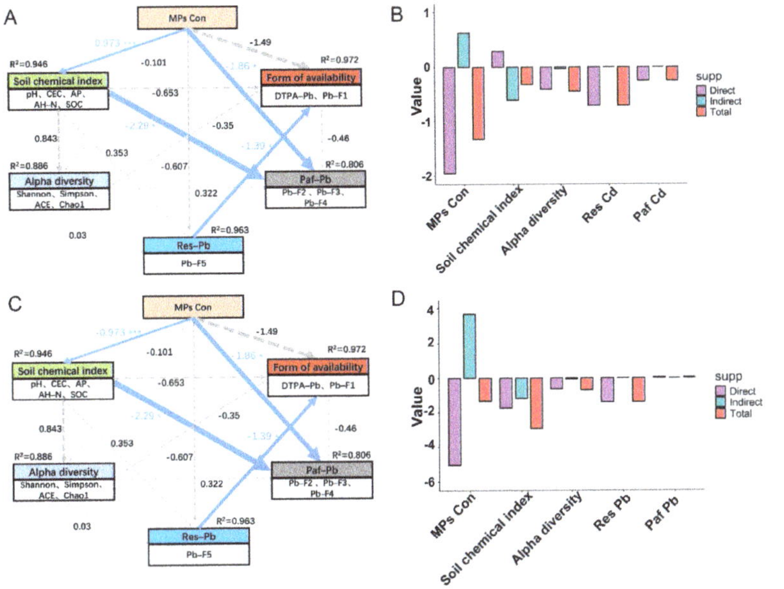

Figure 7. Factors regulating Cd (**A**) and Pb (**C**) availability under the influence of microplastic concentration. PLS–PM illustrates both the direct and indirect influences of observed or latent variables on the availability of Cd and Pb. Each box denotes a collection of observed or latent variables. Positive and negative causal flows are represented by blue and red arrows, respectively. Arrows with increased thickness indicate higher path coefficients. The line width reflects the size of the standardized coefficients (* $p < 0.05$, *** $p < 0.001$). The R^2 value accompanying a parameter reflects the variance explained by other parameters. (**B**) Cd and (**D**) Pb Illustrates the distinction between direct, indirect, and total standardized effects related to the five main variables.

3.7. Analysis of Partial Least Squares Path Modeling

PLS–PM was utilized to evaluate both the direct and indirect effects of PE MP concentration, soil properties, bacterial community alpha–diversity, and the potential bioavailable form (Paf) and residual form (Res) of Cd and Pb on their availability (Figure 7). The PLS–PM analysis revealed that PE MP concentrations had a direct impact on the availability of Cd and Pb, with β coefficients of −1.33 and −1.35, respectively, and significantly influenced the physicochemical properties of the soil ($p < 0.001$). Furthermore, microbial diversity was found to significantly affect DTPA–Cd and the exchangeable forms of Cd (Cd–F1) (β = −0.48, $p < 0.05$). Additionally, the potential bioavailable Cd significantly impacted both DTPA–Cd and Cd–F1 (β = −0.254, $p < 0.05$). The concentration of PE MPs and the physicochemical properties of the soil were found to directly influence the potential bioavailable Pb (Paf–Pb), while the residue–bound Pb significantly affected the bioavailable Pb (β = −1.39, $p < 0.05$).

4. Discussion

4.1. PE MPs Altered Properties of the Soil

Soil pH is recognized as a crucial factor influencing the speciation of metals, their solubility from mineral surfaces, and their mobility and final availability [30]. This study revealed that soil pH decreases with increasing concentrations of PE MPs. Furthermore, soil pH exhibited a tendency to decline gradually over the incubation period (Figure 3A). This phenomenon may be linked to the emission of substances such as aldehydes, esters, and carboxylic acids during the degradation and aging of MPs, which could potentially influence pH levels [31,32]. Additionally, the presence of PE MPs may affect the population of ammonia–oxidizing bacteria and the nitrification process, leading to the release of hydrogen ions and a subsequent decrease in soil pH [33]. The CEC refers to the amount of cations that can be absorbed and exchanged by the soil colloids. An increase in CEC enhances ion exchange processes when soluble metal salts are adsorbed [34,35]. This study found that as the concentration of PE MPs in the soil increases, there is a corresponding gradual decline in soil CEC (Figure 3B). MPs may compete with cations for active sites on clay minerals and soil organic matter, thereby reducing CEC [36]. Furthermore, MPs may exert adsorption or desorption effects on soil organic matter, altering the structure and function of soil microbial communities. This, in turn, influences the decomposition and transformation processes of organic matter, indirectly affecting CEC [37]. Multiple mechanisms collectively determine variations in soil CEC. In the current research, the SOC content increased with rising concentrations of PE MPs during the incubation phase (Figure 3C). These results indicate that PE MPs can act as a source of SOC in the soil. Previous investigations have demonstrated that when the concentration of MPs in the soil reaches 28%, there is a significant increase in DOC levels [17,38]. Importantly, TOC and DOC present in the soil can form stable complexes with HMs. The increased SOC content in the soil further enhances the proportion of metals associated with organic matter (F4), thereby reducing the availability of HMs [39]. This conclusion is consistent with our findings (Figure 2).

4.2. Soil Bacterial Community Responses to Combined Exposure of PE MPs and Cd/Pb

MPs generally impact the soil microbial community through various direct and indirect mechanisms, including the release of pollutants and modifications to soil properties [40]. In this investigation, the Simpson, ACE, and Chao1 indices of the bacterial community increased under low concentrations of PE MPs (0.1% and 1%) treatments, while the addition of a high concentration (10%) of PE MPs led to a decrease in these indices (Figure 4B). The introduction of 10% PE MPs resulted in an enhanced relative abundance of *Proteobacteria*, while simultaneously causing a reduction in the relative abundance of Acidobacteriota. The abundance of *Actinobacteria* and *Chloroflexi* initially decreased and then increased with rising PE MP concentrations (Figure S6). Previous research suggests that *Proteobacteria* participate in the breakdown of PE MPs and can facilitate this process to some extent [41]. Additionally, both *Proteobacteria* and *Chloroflexi* possess genes related to HM resistance, enhancing their survival in HM–contaminated environments [42]. Recent research indicates that exposure to MPs reduces the relative abundance of the *Acidobacteriota* phylum [43]. This effect may be attributed to unstable carbon sources, given that some microbial groups in *Acidobacteriota* are more adapted to low–carbon settings. The addition of MPs may disrupt the balance of these ecosystems [44], potentially clarifying the reduction in *Acidobacteria* abundance noted in this research. A significant positive correlation has been observed between high–concentration PE MPs (7%) treatment and the relative abundance of *Actinobacteria*. This suggests that *Actinobacteria* may play a role in the degradation of MPs [45]. Previous research has demonstrated that under HM stress, HM–sensitive microorganisms tend to decline gradually, while tolerant species become more prevalent [46]. A substantial decrease in the levels of available Cd and Pb was observed at low concentrations of PE MPs (0.1% and 1%) (Figure 1), potentially mitigating the stress imposed by HMs on microorganisms. Conversely, at a 10% concentration of PE MPs, an increase in available Cd and Pb occurred, exacerbating the combined effects of PE MPs and Cd/Pb on the bacterial community. Mantel test analysis indicated a significant positive correlation ($p < 0.01$) between Actinobacteria and various forms of Cd and Pb. A previous study highlighted a favorable relationship between the presence of *Actinobacteria* in soil and the concentrations of copper (Cu), zinc (Zn), lead (Pb), and mercury (Hg) [47]. This relationship could be associated with the various HM resistance genes found in *Actinobacteria*, which assist in mitigating stress caused by HMs [48]. Random forest analysis revealed that the genera *Ardenticatenaceae*, *Marinococcaceae*, and *Obscuribacteraceae* significantly contributed to the bioavailable forms of Cd and Pb (Figure 5C,F). Pearson correlation analysis indicated that *Marinococcaceae* had a significant negative relationship with DTPA–Pb and DTPA–Cd ($p < 0.05$), while *Ardenticatenaceae* and *Obscuribacteraceae* showed significant correlations with soil pH ($p < 0.05$). These findings imply that these bacteria may have substantial implications for the behavior and distribution of Cd and Pb in the soil.

4.3. Effects of PE MPs on the Bioavailability and Distribution of Cd and Pb

0.1% and 1% PE MPs led to a significant decrease in both DTPA–Cd and DTPA–Pb concentrations in the soil (Figure 1). Several potential explanations for these findings are as follows: (1) MPs possess a hydrophobic carbon backbone with limited surface polarity, but environmental weathering introduces oxygen–containing functional groups (e.g., –OH, –C=O, –COOH). These groups provide negatively charged and polar sites that can bind cationic metals through electrostatic attraction and surface complexation [49]. As shown in Figure S8, PE MPs can adsorb a certain amount of Cd(II) and Pb(II), which may be attributed to the PE MPs themselves possessing a certain specific surface area and surface functional groups [50]. Additionally, the addition of PE MP reduces the adsorption of Cd(II) and Pb(II) by the soil compared to pure soil. This phenomenon may be attributed to the dilution effect of MPs and their occupation of active adsorption sites within the soil, thereby reducing the overall adsorption capacity of the soil [29,51]. (2) MPs may integrate into SOC through the adsorption of organic molecules, thereby enhancing the organically bound fraction of HMs in the soil. These interactions may facilitate the conversion of HMs from bioavailable states into forms that are organically bound [52,53], theoretically reducing the concentration of free HM ions in the soil and decreasing their bioavailability. Conversely, the addition of 10% PE MPs increased the DTPA–Cd and DTPA–Pb content in this study, which contradicts previous findings that indicated the addition of 10% PE MPs decreases the availability of Cd [23]. This may be attributed to the reduction in soil pH caused by 10% PE MPs, which decreased by 1.14 units after 90 days of cultivation compared to the CK (Figure 3A). Research indicates that a decrease in soil pH readily facilitates the conversion of Cd/Pb bound in carbonate complexes and Fe–Mn oxide complexes to ion–exchangeable forms [54,55]. Furthermore, Mantel tests and redundancy analysis provided additional confirmation that pH values exhibit a positive correlation with carbonate binding and Fe–Mn oxide binding (Figure 5).

The Mantel test analysis (Figure 5A,D) revealed a significant positive relationship between SOC content and Cd–F4 and Pb–F4 ($p < 0.001$). These results are consistent with previous research conducted by Yu et al. [53]. Random forest analysis indicated that Cd–F1 and Cd–F3 were significant contributors to DTPA–Cd, while Pb–F1, Pb–F2, and Pb–F3 were major contributors to DTPA–Pb (Figure 5C,F). RDA corroborated these findings (Figure 5B,E). Pb–F1 exhibited a positive correlation with DTPA–Pb, whereas Pb–F2 and Pb–F3 displayed negative correlations with DTPA–Pb. The results of PLS–PM further indicated that changes in PE MP concentrations could not only directly affect bioavailable Cd and Pb by altering soil physicochemical properties but also indirectly influence the forms of Cd and Pb through modifications in soil microbial communities (Figure 7). Prior research has demonstrated that MPs alter the soil physicochemical properties, leading to subsequent changes in microbial communities [56]. Various bacteria and fungi in the soil can utilize MPs or their intermediates as a carbon source for degradation [57]. Additionally, microorganisms can colonize the surface of polyurethane MPs, serving as a habitat [58], which further impacts microbial diversity. Variations in the structure of microbial communities induced by MPs could subsequently affect nutrient cycling and the accessibility of Cd in the soil [22,58]. To further understand the transformation mechanism of PE MPs affecting soil Cd and Pb, we characterized the changes in molecular groups on the surface of soil particles using FTIR, XRD, and XPS. Different doses of PE MPs resulted in varying changes in the molecular groups of the soil. FTIR patterns indicated that 10% concentration of PE MPs significantly affects the Si–O–Si and Fe–O linkages (Figure 6A). According to the XPS spectra, both Fe 2p and Si 2p peak intensities decreased, with the lowest atomic percentages observed under the 10% PE MPs treatment (Figure 6C, Table S2). This suggested that PE MPs influenced the chemical forms of Fe and Si. Previous research has indicated a strong correlation between the presence of iron oxides in soil and the accessibility of HMs. For example, the distances between OH–OH groups in iron oxides closely match the geometry of coordination polyhedra with HMs, thereby contributing to the adsorption of cations and anions at various surface locations [59]. Furthermore, the increase in the number of oxygenated and silica–containing functional groups can provide additional adsorption sites, facilitating the complexation of Pb and Cd [60].

Characteristic diffraction peaks were identified through XRD analysis of soil samples at $2\theta = 22.08°$ and $2\theta = 30.96°$, corresponding to the precipitation of PbO and $PbCO_3/Cd(OH)_2$. However, the intensity of these diffraction peaks was diminished following a 10% PE MPs treatment. These findings suggest that a high concentration of PE MPs inhibits the oxidation–precipitation processes during Cd^{2+} stabilization while promoting the formation of PbO, $PbCO_3$, and $Cd(OH)_2$, thereby increasing the solubility and mobility of Cd^{2+} and Pb^{2+}. Furthermore, XPS analysis indicated that the C1s spectrum exhibited a shift in the absorption peak for the 10% PE MPs treatment, changing from 288.7 eV to 278.5 eV in the binding energy compared to the CK. This shift implies an interaction between HMs and the oxygen–containing functional groups present on the soil surface [61]. Notably, low doses of PE MPs (0.1% and 1%) reduced the intensities of the O=C–O and C–O peaks compared to the CK, while simultaneously enhancing the intensity of the –OH peaks (Figure 6E). This phenomenon may be attributed to the 0.1% and 1% PE MPs enhancing the adhesion of Cd and Pb to the soil, resulting in a decrease in the intensity and presence of O=C–O and C–O peaks, while concurrently increasing both the height and proportion of the O–H peaks. This suggests that metal cations, which are electron–deficient, diminish the electron cloud density surrounding these functional groups, thereby promoting the fixation of metals [62]. In contrast, a 10% concentration of PE MPs resulted in an increase in the C–O and O=C–O peaks, while the –OH peaks decreased in the high–resolution O1s spectra when compared to lower concentrations of PE MPs. This observation may be attributed to the substantial coverage of active sites related to reactive oxygen by MPs, which significantly diminishes the chemisorption capacity of soil for Cd [29].

5. Conclusions

This study revealed that PE MPs influence the bioavailability of soil Cd and Pb through a concentration–dependent effect and elucidated its mechanism. The results indicated that the presence of 0.1% and 1% PE MPs resulted in a reduction in DTPA–extractable Cd and Pb in the soil, while 10% PE MPs increased them. Lower levels of PE MPs (0.1% and 1%) effectively minimized the exchangeable forms of Cd and Pb, thereby facilitating their transformation into more stable compounds, specifically those bound to carbonate, Fe–Mn oxides, and organic matter. Conversely, the higher concentration of 10% PE MPs was found to decrease the fractions of Cd and Pb associated with carbonate and Fe–Mn oxides, while enhancing the exchangeable and residual fractions. The diversity and structure of soil bacterial communities undergo changes under varying PE MP concentrations, with high concentrations exerting a detrimental effect. Correlation analysis revealed strong associations between soil pH, CEC, and SOC content, forms of Cd and Pb, and the bacterial community. FTIR, XRD, and XPS analyses revealed that 0.1% and 1% PE MPs promoted the surface fixation of Cd and Pb with oxygen–containing functional groups, whereas 10% PE MPs inhibited this process and disrupted the formation of PbO, $PbCO_3$, and $Cd(OH)_2$, thereby increasing the availability of Cd in the soil. Random forest modeling indicated that AP and SOC were the critical factors determining DTPA–Cd and Pb, respectively, and the concentration of PE MPs also significantly affected both DTPA–Cd and Pb. Partial least squares path modeling demonstrated that PE MPs altered the soil physicochemical characteristics and the structure of the bacterial community, ultimately impacting the transformation of Cd and Pb speciation, and these changes were highly dependent on PE MP concentration. This study underscores the intricate and diverse impacts of PE MP pollution on soil ecosystems, and the research elucidates how different concentrations of PE MPs affect the availability of Cd and Pb in the soil environment, and provides further insights into the treatment of Cd and Pb–contaminated soils containing MPs.

Supplementary Materials: The following supporting information can be downloaded at: https://www.mdpi.com/article/10.3390/toxics13100901/s1, Supplementary materials Text S1: The detailed procedure for the determination of 5 forms of Cd and Pb; Figure S1: Scanning electron microscope (SEM) images of PE MPs (A), FTIR spectra images of PE MPs (B); Figure S2: Changes in pH in the soil at varying PE MP concentrations. Different letters indicate the significant differences ($p < 0.05$); Figure S3: Changes in AP (A) and AH–N (B) concentrations in the soil over time at varying PE MP concentrations. Different letters indicate the significant differences at the same time point. ($p < 0.05$); Figure S4: Venn diagrams analysis of soil with different PE MP concentration; Figure S5: A: The 10 most important bacterial genera in the soil predicted by random forest models. B: Differences in the relative abundance of the top ten bacterial genera ranked by importance across treatments; Figure S6: The relative abundance of bacterial community in soil at the phylum level; Figure S7: Correlations between soil most important bacterial genera and soil properties. Significant levels are indicated: * $p < 0.05$, ** $p < 0.01$, *** $p < 0.001$; Figure S8: Adsorption of Cd and Pb by PE MPs soil and soil + PE MPs. (The adsorption conditions were set as follows: adsorbent amount = 0.1 g, the concentration of the Cd/Pb solution = 5 mg/L. V = 25 mL, pH 6.8, equilibrium time 1440 min); Table S1: The physicochemical properties of the test soil; Table S2: Atomic percentages (%) for the elements on the sample particle surfaces.

Author Contributions: Conceptualization, Z.W.; Methodology, Z.W.; Formal analysis, Z.W., S.L., G.L. and R.D.; Investigation, S.L., G.L. and R.D.; Data curation, Z.W.; Writing—original draft, Z.W.; Writing—review & editing, Z.W., P.Z., C.L. and H.F.; Visualization, Z.W., S.L., G.L. and R.D.; Supervision, P.Z. and H.F.; Project administration, P.Z. and H.F.; Funding acquisition, P.Z. and H.F. All authors have read and agreed to the published version of the manuscript.

Funding: This research was funded by the national key research and development program of China (grant number 2021YFD1700900), Science and Technology Innovation Fund of Henan Agricultural University (grant number 2024CXZX012), Scientific and Technological Project of Henan Province (grant number 242102320137), and Special fund for young talents in Henan Agricultural University (grant number 30501356).

Institutional Review Board Statement: Not applicable.

Informed Consent Statement: Not applicable.

Data Availability Statement: The data of this study are available on request from the corresponding author.

Acknowledgments: We are grateful to editors and reviewers for their meaningful suggestions.

Conflicts of Interest: The authors declare no conflicts of interest.

References

1. Duan, C.; Fang, L.; Yang, C.; Chen, W.; Cui, Y.; Li, S. Reveal the response of enzyme activities to heavy metals through in situ zymography. *Ecotoxicol. Environ. Saf.* **2018**, *156*, 106–115. [CrossRef]
2. Yu, Y.; Rong, K.; Sui, X.; Zhang, L.; Zhang, M.; Hu, H.; Jia, J.; Wu, J.; Li, C. Analysis of *NRAMP* genes in the Triticeae reveals that *TaNRAMP5* positively regulates cadmium (Cd) tolerance in wheat (*Triticum aestivum*). *Plant Physiol. Biochem.* **2025**, *219*, 109321. [CrossRef]
3. Li, C.; Li, G.; Wang, Y.; Wang, J.; Liu, H.; Gao, W.; Qin, S.; Sui, F.; Fu, H.; Zhao, P. Supplementing two wheat genotypes with $ZnSO_4$ and ZnO nanoparticles showed differential mitigation of Cd phytotoxicity by reducing Cd absorption, preserving root cellular ultrastructure, and regulating metal-transporter gene expression. *Plant Physiol. Biochem.* **2024**, *206*, 108199. [CrossRef] [PubMed]
4. Dawar, K.; Mian, I.A.; Khan, S.; Zaman, A.; Danish, S.; Liu, K.; Harrison, M.T.; Saud, S.; Hassan, S.; Nawaz, T.; et al. Alleviation of cadmium toxicity and fortification of zinc in wheat cultivars cultivated in Cd contaminated soil. *S. Afr. J. Bot.* **2023**, *162*, 611–621. [CrossRef]
5. Wu, B.; Liu, Z.; Xu, Y.; Li, D.; Li, M. Combined toxicity of cadmium and lead on the earthworm *Eisenia fetida* (Annelida, Oligochaeta). *Ecotoxicol. Environ. Saf.* **2012**, *81*, 122–126. [CrossRef] [PubMed]
6. Wei, B.; Yang, L. A review of heavy metal contaminations in urban soils, urban road dusts and agricultural soils from China. *Microchem. J.* **2010**, *94*, 99–107. [CrossRef]
7. Qiang, L.; Hu, H.; Li, G.; Xu, J.; Cheng, J.; Wang, J.; Zhang, R. Plastic mulching, and occurrence, incorporation, degradation, and impacts of polyethylene microplastics in agroecosystems. *Ecotoxicol. Environ. Saf.* **2023**, *263*, 115274. [CrossRef]
8. Thompson, R.C.; Olsen, Y.; Mitchell, R.P.; Davis, A.; Rowland, S.J.; John, A.W.G.; McGonigle, D.; Russell, A.E. Lost at sea: Where is all the plastic? *Science* **2004**, *304*, 838. [CrossRef]
9. Zhao, S.; Zhang, Z.; Chen, L.; Cui, Q.; Cui, Y.; Song, D.; Fang, L. Review on migration, transformation and ecological impacts of microplastics in soil. *Appl. Soil Ecol.* **2022**, *176*, 104486. [CrossRef]
10. Kang, Q.; Zhang, K.; Dekker, S.C.; Mao, J. Microplastics in soils: A comprehensive review. *Sci. Total Environ.* **2025**, *960*, 178298. [CrossRef]
11. Liu, X.; Shao, J.; Peng, C.; Gong, J. Novel insights related to soil microplastic abundance and vegetable microplastic contamination. *J. Hazard. Mater.* **2025**, *484*, 136727. [CrossRef]
12. An, Q.; Zhou, T.; Wen, C.; Yan, C. The effects of microplastics on heavy metals bioavailability in soils: A meta-analysis. *J. Hazard. Mater.* **2023**, *460*, 132369. [CrossRef] [PubMed]
13. Sungur, A.; Soylak, M.; Ozcan, H. Investigation of heavy metal mobility and availability by the BCR sequential extraction procedure: Relationship between soil properties and heavy metals availability. *Chem. Spec. Bioavailab.* **2014**, *26*, 219–230. [CrossRef]
14. Wang, F.; Zhang, X.; Zhang, S.; Zhang, S.; Adams, C.A.; Sun, Y. Effects of Co-Contamination of Microplastics and Cd on Plant Growth and Cd Accumulation. *Toxics* **2020**, *8*, 36. [CrossRef]
15. Halim, M.; Conte, P.; Piccolo, A. Potential availability of heavy metals to phytoextraction from contaminated soils induced by exogenous humic substances. *Chemosphere* **2003**, *52*, 265–275. [CrossRef]
16. Semenov, V.M.; Tulina, A.S.; Semenova, N.A.; Ivannikova, L.A. Humification and Nonhumification Pathways of the Organic Matter Stabilization in Soil: A Review. *Eurasian Soil Sci.* **2013**, *46*, 355–368. [CrossRef]
17. Wang, F.; Zhang, X.; Zhang, S.; Zhang, S.; Sun, Y. Interactions of microplastics and cadmium on plant growth and arbuscular mycorrhizal fungal communities in an agricultural soil. *Chemosphere* **2020**, *254*, 126791. [CrossRef] [PubMed]
18. Meng, Q.; Diao, T.; Yan, L.; Sun, Y. Effects of single and combined contamination of microplastics and cadmium on soil organic carbon and microbial community structural: A comparison with different types of soil. *Appl. Soil Ecol.* **2023**, *183*, 104763. [CrossRef]
19. Zeb, A.; Liu, W.; Meng, L.; Lian, J.; Wang, Q.; Lian, Y.; Chen, C.; Wu, J. Effects of polyester microfibers (PMFs) and cadmium on lettuce (*Lactuca sativa*) and the rhizospheric microbial communities: A study involving physio-biochemical properties and metabolomic profiles. *J. Hazard. Mater.* **2022**, *424*, 127405. [CrossRef]
20. Tang, Y.; Xing, Y.; Wang, X.; Ya, H.; Zhang, T.; Lv, M.; Wang, J.; Zhang, H.; Dai, W.; Zhang, D.; et al. PET microplastics influenced microbial community and heavy metal speciation in heavy-metal contaminated soils. *Appl. Soil Ecol.* **2024**, *201*, 105488. [CrossRef]
21. Jiang, X.; Yang, Y.; Wang, Q.; Liu, N.; Li, M. Seasonal variations and feedback from microplastics and cadmium on soil organisms in agricultural fields. *Environ. Int.* **2022**, *161*, 107096. [CrossRef] [PubMed]
22. Feng, X.; Wang, Q.; Sun, Y.; Zhang, S.; Wang, F. Microplastics change soil properties, heavy metal availability and bacterial community in a Pb-Zn-contaminated soil. *J. Hazard. Mater.* **2022**, *424*, 127364. [CrossRef] [PubMed]
23. Li, C.; Sun, H.; Shi, Y.; Zhao, Z.; Zhang, Z.; Zhao, P.; Gao, Q.; Zhang, X.; Chen, B.; Li, Y.; et al. Polyethylene and poly (butyleneadipate-co-terephthalate)-based biodegradable microplastics modulate the bioavailability and speciation of Cd and As in soil: Insights into transformation mechanisms. *J. Hazard. Mater.* **2023**, *445*, 130638. [CrossRef] [PubMed]

24. Khan, A.B.; Pereao, O.; Sparks, C.; Opeolu, B. Assessing microplastic characteristics and abundance in the sediment and surface water of the Diep River, Western Cape, South Africa. *Environ. Pollut.* **2025**, *381*, 126555. [CrossRef]
25. Li, Q.; Yan, J.; Li, Y.; Liu, Y.; Andom, O.; Li, Z. Microplastics alter cadmium accumulation in different soil-plant systems: Revealing the crucial roles of soil bacteria and metabolism. *J. Hazard. Mater.* **2024**, *474*, 134768. [CrossRef]
26. Niu, R.; Zhu, C.; Jiang, G.; Yang, J.; Zhu, X.; Li, L.; Shen, F.; Jie, X.; Liu, S. Variations in Soil Nitrogen Availability and Crop Yields under a Three-Year Annual Wheat and Maize Rotation in a Fluvo-Aquic Soil. *Plants* **2023**, *12*, 808. [CrossRef]
27. Tessier, A.; Campbell, P.G.C.; Bisson, M. Sequential extraction procedure for the speciation of particulate trace metals. *Anal. Chem.* **1979**, *51*, 844–851. [CrossRef]
28. Zhu, Z.; Sun, L.; Qin, Q.; Sun, Y.; Yang, S.; Wang, J.; Yang, Y.; Gao, G.; Xue, Y. The Adsorption Process and Mechanism of Benzo[a]pyrene in Agricultural Soil Mediated by Microplastics. *Toxics* **2024**, *12*, 922. [CrossRef]
29. Guo, J.-J.; Li, F.; Xiao, H.-C.; Liu, B.-L.; Feng, L.-N.; Yu, P.-F.; Meng, C.; Zhao, H.-M.; Feng, N.-X.; Li, Y.-W.; et al. Polyethylene and polypropylene microplastics reduce chemisorption of cadmium in paddy soil and increase its bioaccessibility and bioavailability. *J. Hazard. Mater.* **2023**, *449*, 130994. [CrossRef]
30. Zhao, K.; Liu, X.; Xu, J.; Selim, H.M. Heavy metal contaminations in a soil-rice system: Identification of spatial dependence in relation to soil properties of paddy fields. *J. Hazard. Mater.* **2010**, *181*, 778–787. [CrossRef]
31. Bandow, N.; Will, V.; Wachtendorf, V.; Simon, F.-G. Contaminant release from aged microplastic. *Environ. Chem.* **2017**, *14*, 394–405. [CrossRef]
32. Binda, G.; Kalcikova, G.; Allan, I.J.; Hurley, R.; Rodland, E.; Spanu, D.; Nizzetto, L. Microplastic aging processes: Environmental relevance and analytical implications. *TrAC Trends Anal. Chem.* **2024**, *172*, 117566. [CrossRef]
33. Rong, L.; Zhao, L.; Zhao, L.; Cheng, Z.; Yao, Y.; Yuan, C.; Wang, L.; Sun, H. LDPE microplastics affect soil microbial communities and nitrogen cycling. *Sci. Total Environ.* **2021**, *773*, 145640. [CrossRef]
34. Hong, M.; Yu, L.; Wang, Y.; Zhang, J.; Chen, Z.; Dong, L.; Zan, Q.; Li, R. Heavy metal adsorption with zeolites: The role of hierarchical pore architecture. *Chem. Eng. J.* **2019**, *359*, 363–372. [CrossRef]
35. Ji, M.; Wang, X.; Usman, M.; Liu, F.; Dan, Y.; Zhou, L.; Campanaro, S.; Luo, G.; Sang, W. Effects of different feedstocks-based biochar on soil remediation: A review. *Environ. Pollut.* **2022**, *294*, 118655. [CrossRef]
36. Peng, H.; Lin, Z.; Lu, D.; Yu, B.; Li, H.; Yao, J. How do polystyrene microplastics affect the adsorption of copper in soil? *Sci. Total Environ.* **2024**, *924*, 171545. [CrossRef]
37. Zhai, Y.; Bai, J.; Chang, P.; Liu, Z.; Wang, Y.; Liu, G.; Cui, B.; Peijnenburg, W.; Vijver, M.G. Microplastics in terrestrial ecosystem: Exploring the menace to the soil-plant-microbe interactions. *TrAC Trends Anal. Chem.* **2024**, *174*, 117667. [CrossRef]
38. Liu, H.; Yang, X.; Liu, G.; Liang, C.; Xue, S.; Chen, H.; Ritsema, C.J.; Geissen, V. Response of soil dissolved organic matter to microplastic addition in Chinese loess soil. *Chemosphere* **2017**, *185*, 907–917. [CrossRef]
39. Liu, S.; Huang, J.; Zhang, W.; Shi, L.; Yi, K.; Yu, H.; Zhang, C.; Li, S.; Li, J. Microplastics as a vehicle of heavy metals in aquatic environments: A review of adsorption factors, mechanisms, and biological effects. *J. Environ. Manag.* **2022**, *302*, 113995. [CrossRef] [PubMed]
40. Qiu, Y.; Zhou, S.; Zhang, C.; Zhou, Y.; Qin, W. Soil microplastic characteristics and the effects on soil properties and biota: A systematic review and meta-analysis. *Environ. Pollut.* **2022**, *313*, 120183. [CrossRef]
41. Wu, C.; Ma, Y.; Wang, D.; Shan, Y.; Song, X.; Hu, H.; Ren, X.; Ma, X.; Cui, J.; Ma, Y. Integrated microbiology and metabolomics analysis reveal plastic mulch film residue affects soil microorganisms and their metabolic functions. *J. Hazard. Mater.* **2022**, *423*, 127258. [CrossRef] [PubMed]
42. Chen, Y.; Jiang, Y.; Huang, H.; Mou, L.; Ru, J.; Zhao, J.; Xiao, S. Long-term and high-concentration heavy-metal contamination strongly influences the microbiome and functional genes in Yellow River sediments. *Sci. Total Environ.* **2018**, *637*, 1400–1412. [CrossRef] [PubMed]
43. Meng, J.; Diao, C.; Cui, Z.; Li, Z.; Zhao, J.; Zhang, H.; Hu, M.; Xu, J.; Jiang, Y.; Haider, G.; et al. Unravelling the influence of microplastics with/without additives on radish (*Raphanus sativus*) and microbiota in two agricultural soils differing in pH. *J. Hazard. Mater.* **2024**, *478*, 135535. [CrossRef]
44. Palansooriya, K.N.; Withana, P.A.; Jeong, Y.; Sang, M.K.; Cho, Y.; Hwang, G.; Chang, S.X.; Ok, Y.S. Contrasting effects of food waste and its biochar on soil properties and lettuce growth in a microplastic-contaminated soil. *Appl. Biol. Chem.* **2024**, *67*, 3. [CrossRef]
45. Palansooriya, K.N.; Shi, L.; Sarkar, B.; Parikh, S.J.; Sang, M.K.; Lee, S.-R.; Ok, Y.S. Effect of LDPE microplastics on chemical properties and microbial communities in soil. *Soil Use Manag.* **2022**, *38*, 1481–1492. [CrossRef]
46. Xiang, Y.; Lan, J.; Dong, Y.; Zhou, M.; Hou, H.; Huang, B.-T. Pollution control performance of solidified nickel-cobalt tailings on site: Bioavailability of heavy metals and microbial response. *J. Hazard. Mater.* **2024**, *471*, 134295. [CrossRef]
47. Sun, T.; Li, G.; Mazarji, M.; Delaplace, P.; Yang, X.; Zhang, J.; Pan, J. Heavy metals drive microbial community assembly process in farmland with long-term biosolids application. *J. Hazard. Mater.* **2024**, *468*, 133845. [CrossRef]

48. Yan, C.; Wang, F.; Geng, H.; Liu, H.; Pu, S.; Tian, Z.; Chen, H.; Zhou, B.; Yuan, R.; Yao, J. Integrating high-throughput sequencing and metagenome analysis to reveal the characteristic and resistance mechanism of microbial community in metal contaminated sediments. *Sci. Total Environ.* **2020**, *707*, 136116. [CrossRef]
49. Holmes, L.A.; Turner, A.; Thompson, R.C. Adsorption of trace metals to plastic resin pellets in the marine environment. *Environ. Pollut.* **2012**, *160*, 42–48. [CrossRef]
50. Kong, X.; Zhou, A.; Chen, X.; Cheng, X.; Lai, Y.; Li, C.; Ji, Q.; Ji, Q.; Kong, J.; Ding, Y.; et al. Insight into the adsorption behaviors and bioaccessibility of three altered microplastics through three types of advanced oxidation processes. *Sci. Total Environ.* **2024**, *917*, 170420. [CrossRef] [PubMed]
51. Zhang, S.; Han, B.; Sun, Y.; Wang, F. Microplastics influence the adsorption and desorption characteristics of Cd in an agricultural soil. *J. Hazard. Mater.* **2020**, *388*, 121775. [CrossRef] [PubMed]
52. Yu, H.; Zhang, Z.; Zhang, Y.; Fan, P.; Xi, B.; Tan, W. Metal type and aggregate microenvironment govern the response sequence of speciation transformation of different heavy metals to microplastics in soil. *Sci. Total Environ.* **2021**, *752*, 141956. [CrossRef]
53. Yu, H.; Hou, J.; Dang, Q.; Cui, D.; Xi, B.; Tan, W. Decrease in bioavailability of soil heavy metals caused by the presence of microplastics varies across aggregate levels. *J. Hazard. Mater.* **2020**, *395*, 122690. [CrossRef]
54. Zeng, F.; Ali, S.; Zhang, H.; Ouyang, Y.; Qiu, B.; Wu, F.; Zhang, G. The influence of pH and organic matter content in paddy soil on heavy metal availability and their uptake by rice plants. *Environ. Pollut.* **2011**, *159*, 84–91. [CrossRef]
55. Shen, B.; Wang, X.; Zhang, Y.; Zhang, M.; Wang, K.; Xie, P.; Ji, H. The optimum pH and Eh for simultaneously minimizing bioavailable cadmium and arsenic contents in soils under the organic fertilizer application. *Sci. Total Environ.* **2020**, *711*, 135229. [CrossRef] [PubMed]
56. Qi, R.; Jones, D.L.; Tang, Y.; Gao, F.; Li, J.; Chi, Y.; Yan, C. Regulatory path for soil microbial communities depends on the type and dose of microplastics. *J. Hazard. Mater.* **2024**, *473*, 134702. [CrossRef] [PubMed]
57. Wang, J.; Peng, C.; Li, H.; Zhang, P.; Liu, X. The impact of microplastic-microbe interactions on animal health and biogeochemical cycles: A mini-review. *Sci. Total Environ.* **2021**, *773*, 145697. [CrossRef]
58. Zhao, M.; Liu, R.; Wang, X.; Zhang, J.; Wang, J.; Cao, B.; Zhao, Y.; Xu, L.; Chen, Y.; Zou, G. How do controlled-release fertilizer coated microplastics dynamically affect Cd availability by regulating Fe species and DOC content in soil? *Sci. Total Environ.* **2022**, *850*, 157886. [CrossRef]
59. Manceau, A.; Charlet, L.; Boisset, M.; Didier, B.; Spadini, L. Sorption and speciation of heavy metals on hydrous Fe and Mn oxides. From microscopic to macroscopic. *Appl. Clay Sci.* **1992**, *7*, 201–223. [CrossRef]
60. Jiang, X.; Liu, Z.; Yan, B.; Zhao, L.; Chen, T.; Yang, X. Effects of active silicon amendment on Pb(II)/Cd(II) adsorption: Performance evaluation and mechanism. *J. Hazard. Mater.* **2024**, *478*, 135614. [CrossRef]
61. Guo, X.; Du, B.; Wei, Q.; Yang, J.; Hu, L.; Yan, L.; Xu, W. Synthesis of amino functionalized magnetic graphenes composite material and its application to remove Cr(VI), Pb(II), Hg(II), Cd(II) and Ni(II) from contaminated water. *J. Hazard. Mater.* **2014**, *278*, 211–220. [CrossRef] [PubMed]
62. Li, Y.; Wang, K.; Dötterl, S.; Xu, J.; Garland, G.; Liu, X. The critical role of organic matter for cadmium-lead interactions in soil: Mechanisms and risks. *J. Hazard. Mater.* **2024**, *476*, 135123. [CrossRef] [PubMed]

toxics

MDPI

Article

Effects of Foliar Spraying of Dicarboxylicdimethylammonium Chloride on Cadmium and Arsenic Accumulation in Rice Grains

Lin Fu [1,†], Jiawei Deng [1,†], Dayliana Ruiz Lao [1,†], Changbo Zhang [1,*], Weijie Xue [1,*], Yun Deng [2] and Xin Luo [1]

1 Key Laboratory of Original Agro-Environmental Pollution Prevention and Control, Ministry of Agriculture and Rural Affairs, Agro-Environmental Protection Institute, Ministry of Agriculture and Rural Affairs, Tianjin 300191, China; fl1004197187@163.com (L.F.); d17736218037@163.com (J.D.); 18501338578@163.com (D.R.L.); l_xin01@163.com (X.L.)

2 School of Environment and Ecology, Jiangnan University, Wuxi 214122, China; dengyun@jiangnan.edu.cn

* Correspondence: zhangchangbo@caas.cn (C.Z.); tzfelicity@163.com (W.X.)

† The authors contributed equally to this work.

Abstract: A field experiment with double cropping rice was carried out to study the foliar application effects of dicarboxylicdimethylammonium chloride (DDAC) on cadmium (Cd) and arsenic (As) accumulation in rice grains. The results showed that the spraying of DDAC could significantly reduce the accumulation of Cd and As in rice grains. The highest reductions in Cd and As content were observed when 1.5 mmol L^{-1} DDAC was sprayed, with 49.1% and 27.4% reductions in Cd and As content in early rice grains and 56.5% and 28.1% reductions in Cd and As content in late rice grains, respectively. In addition, the content of calcium (Ca) in rice grains increased significantly after DDAC foliar application, which was also conducive to the synthesis of amino acids such as glutamate (Glu), glycine (Gly) and cysteine (Cys) in rice grains. The results indicated that the foliar spraying of DDAC can inhibit the absorption, transport, accumulation and toxicity of Cd and As in rice grains by increasing amino acid synthesis and regulating the absorption and transport of essential elements.

Keywords: foliar inhibitor; dicarboxylicdimethylammonium chloride; rice; cadmium; arsenic; accumulation

1. Introduction

As two non-essential-harmful elements, cadmium (Cd) and arsenic (As) are easily absorbed by rice through the soil-plant system and are transported to grains for accumulation [1,2]. Emissions from industry and mining, as well as the misuse of fertilizers and pesticides, lead to widespread Cd and As production and their co-contamination of paddy soil in China, posing a threat to food security and human health [3–5].

Due to the opposite environmental behaviors of Cd and As in soil, the synchronous control of Cd-As composite pollution is extremely challenging [6]. Water management is widely used in the remediation of Cd-As composite pollution, but the effectiveness of joint governance is often unable to be achieved due to the inability to strictly control soil pH and Eh [7,8]. Treatment methods involving the addition of Cd and As passivators to soil often pose the risks of reactivation and secondary pollution [9], and 50–70% of the Cd and As in rice grains mainly comes from the remigration and reactivation of Cd and As accumulated in vegetative organs before the flowering period [10,11]. Therefore, inhibiting the activity of Cd and As in vegetative organs is one of the most effective pathways for the simultaneous remediation of co-contamination with Cd and As, such as in foliar spraying technology, which can inhibit the remigration and reactivation of heavy metals [12,13].

The transport of Cd in plants is usually carried out by metal ion transporters or cation transport channels such as K^+, Ca^{2+}, Mg^{2+}, Mn^{2+}, Fe^{2+}, and Zn^{2+}, which leads to a competitive relationship between Cd and these cations [14]. The exogenous addition of K, Ca, Mn, Fe, or Zn significantly increases the absorption and transport of essential cations by rice, while reducing the absorption and transport of Cd [15]. The absorption pathway of As in rice is related to its current form, which can be divided into silicate (Si) transporters and phosphate (P) transporters, leading to competition between As and Si and P [16]. Essential elements are necessary for plant metabolism because they act as enzyme regulators, cofactors, and activators [17]. Furthermore, the transport rate of beneficial elements in plants is much higher than that of harmful elements [6]. Therefore, increasing the absorption of essential elements that have antagonistic effects on Cd and As may be an important method of inhibiting the absorption of Cd and As by rice.

Amino acids are the basic units of proteins, as well as the precursors of small molecules such as antioxidants and signaling metabolites, which are closely related to plant growth metabolism, the abiotic stress response, and element absorption and transport [18–20]. As the central substance of amino acid metabolism, glutamate (Glu) plays an important role in alleviating the toxicity of Cd and As. Studies have shown that the exogenous application of Glu downregulates the expression levels of Cd transporter-related genes in roots and alleviates Cd-induced chlorosis and growth inhibition, which demonstrates that any pathway that increases endogenous Glu content has the potential to reduce heavy metal uptake in rice and to alleviate heavy metal toxicity [21]. The exogenous application of Glu to rice under As stress protects photosynthetic function and the growth of rice plants by affecting nitrogen assimilation, proline metabolism, and the antioxidant system [22]. Unfortunately, the water solubility of Glu is extremely low, so direct foliar spraying of Glu has no significant effect on the endogenous Glu content in rice.

In order to increase the water solubility of Glu, an ionic liquid named dicarboxylicdimethylammonium chloride (DDAC) was synthesized using Glu as the raw material, which has the characteristics of good environmental protection, strong water solubility, and high stability (Scheme 1). Ionic liquids with glycine as a precursor have been shown to effectively reduce Cd accumulation in rice seedlings [23,24]. However, there are no reports on the effect of DDAC on the regulation of Cd and As accumulation in rice. Therefore, a paddy field experiment was conducted to explore the effects of the foliar application of DDAC on the content of Cd, As, essential elements, and total amino acids in rice grains under Cd-As co-contamination. The results showed that DDAC could effectively reduce the accumulation of Cd and As in rice grains by regulating the absorption and transport of essential elements and promoting the synthesis of amino acids.

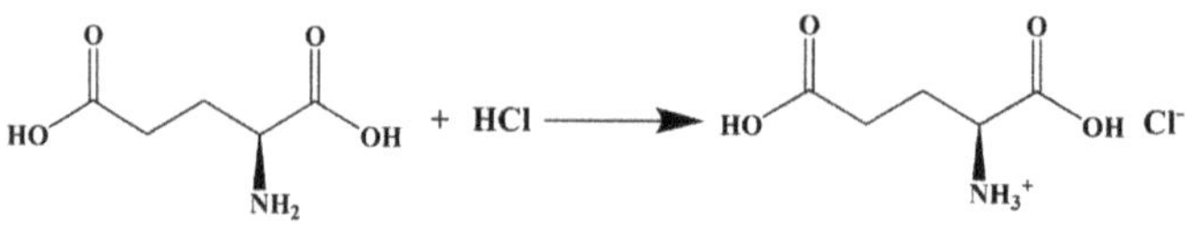

Scheme 1. Synthetic form of DDAC.

2. Materials and Methods

2.1. Material Preparation

In hydrochloric acid solution, the same molar number of glutamate was added and stirred evenly. The liquid was transferred to a round-bottom flask after being kept at a constant temperature reaction at 60 °C for 2 h, and was then distilled under reduced pressure with a rotary evaporator to obtain a white solid powder at room temperature—that is, DDAC.

2.2. Experimental Design

The field trial was conducted in Xiangtan City, Hunan Province of China (N: 27°36′, E: 112°58′), with two seasons of early rice and late rice. The test soil was Cd-As co-contaminated soil with a Cd content of 1.29 mg kg^{-1} and an As content of 39.29 mg kg^{-1}. Early rice was sown in late April, sprayed in mid-June, and harvested in mid-late July. Late rice was sown in late July, sprayed in mid to late September, and harvested in November. The experiment was designed as a randomized complete block design with four replications for each treatment. Based on the results of preliminary studies and tests, different concentrations of DDAC solutions were sprayed at the flowering stage of the rice as follows: CK (control group), T1 (0.20 mmol L^{-1}), T2 (0.50 mmol L^{-1}), T3 (0.80 mmol L^{-1}), T4 (1.20 mmol L^{-1}), and T5 (1.50 mmol L^{-1}) [24]. Each spray treatment was performed once a day for two consecutive days. Field management was basically consistent with that of local high-yield rice fields. Samples were collected from rice plants at maturity and dried to determine Cd, As, essential element and amino acid levels in the grains.

2.3. Determination of Cd and As Content

The determination of Cd and As content was completed according to previous reports [10,25]. After accurately weighing 0.5 g of the rice grains into the digestion tube, 7.0 mL of MOS HNO_3 was added and left for more than 5 h. DigiBlock ED54 (LabTech, Beijing, China) was used for digestion at 110 °C for 2.5 h. After cooling to room temperature, 1.0 mL of H_2O_2 was added to continue the digestion process for 1.5 h. Finally, the temperature was raised to 170 °C for acid removal. Deionized water was added to the digestion solution to 25.0 mL, which was then filtered for the detection of Cd content. Inductively coupled plasma mass spectrometry iCAP Q ICP-MS (Thermo Scientific, Waltham, MA, USA) was used for Cd determination. The processing method of As determination was similar to that of Cd, except that the digestion temperature was fixed at 110 °C for 4 h and H_2O_2 was no longer added. An atomic fluorescence spectrometer AFS-8520 (Haiguang, Beijing, China) was used to determine the As content. Standard reference material (Rice Powder Certified Reference Material GBW(E)100350) and blank digestion samples were used for quality assurance and quality controls (QA/QCs). The recovery rate was 90–105% to ensure the accuracy and trustworthiness of the data.

2.4. Determination of Essential Element Content

The determination of the essential element content was completed according to previous reports [26]. ICP-MS was also used to determine the content of K, Ca, Mg, Fe, Mn and Zn, and the sample treatment method was the same as in Section 2.2.

2.5. Determination of Total Amino Acid Content

The extraction and identification of amino acids were completed on the basis of previous reports [27]. After accurately weighing 0.25 g grain powder in a test tube, 15.0 mL of 6 mol L^{-1} HCl was added. The test tube was placed on ice to cool for 5 min, filled with high-purity nitrogen, sealed, and placed in a 110 °C constant-temperature air drying oven for 22 h. The acid solution was cooled to room temperature and filtered, and then deionized water was added to 50.0 mL; 1.0 mL of the above solution was transferred to a clean test tube and dried with nitrogen in a 50 °C water bath. The steps were repeated with 1.0 mL of deionized water. Finally, the precipitate was dissolved in sodium citrate buffer (pH = 2.2) and filtered with a 0.22 μm filter membrane (JIN TENG, Tianjin, China) for detection. The total amino acid content was determined using an Agilent 1200 high-performance liquid chromatograph (Agilent Technologies, Palo Alto, CA, USA). An advance Bio AAA column (100 mm × 4.6 mm, 2.7 μm, Agilent Technologies, USA) and diode array detector (DAD) were used.

2.6. Statistical Analysis

All data were expressed as mean ± standard deviation. One-way analysis of variance (ANOVA) and Duncan's tests were used to analyze the significant differences between treatments ($p < 0.05$). Microsoft Excel 2021 and SPSS 26.0 were used for data collation and analysis and Origin 2021 was used for data visualization.

3. Results

3.1. Effects of DDAC on the Acumulation of Cd and As in Rice Grains

The content of Cd and As in the rice grains decreased significantly after foliar spraying of DDAC at flowering stage (Figure 1). Compared with CK, the Cd content in early rice grains decreased from 0.31 mg kg^{-1} to 0.22–0.16 mg kg^{-1}, with a decrease of 28.8–49.1% when DDAC was sprayed at different concentrations. The As content decreased from 0.35 mg kg^{-1} to 0.30–0.26 mg kg^{-1} by 14.0–27.4%. The accumulation of Cd and As in late rice was higher than that in early rice. In the control group, the Cd and As content in late rice grains was 0.36 mg kg^{-1} and 0.38 mg kg^{-1}, respectively. After the foliar spraying of DDAC, the Cd content decreased to 0.21–0.16 mg kg^{-1}, with a decrease of 40.9–56.5%, and the As content decreased to 0.32–0.27 mg kg^{-1}, with a decrease of 15.4–28.1%. In summary, DDAC, as a foliar inhibitor, can effectively alleviate both Cd and As accumulation in rice grains in Cd-As-contaminated areas. In addition, the mitigation effect of DDAC on Cd and As accumulation generally increased with increases in its concentration.

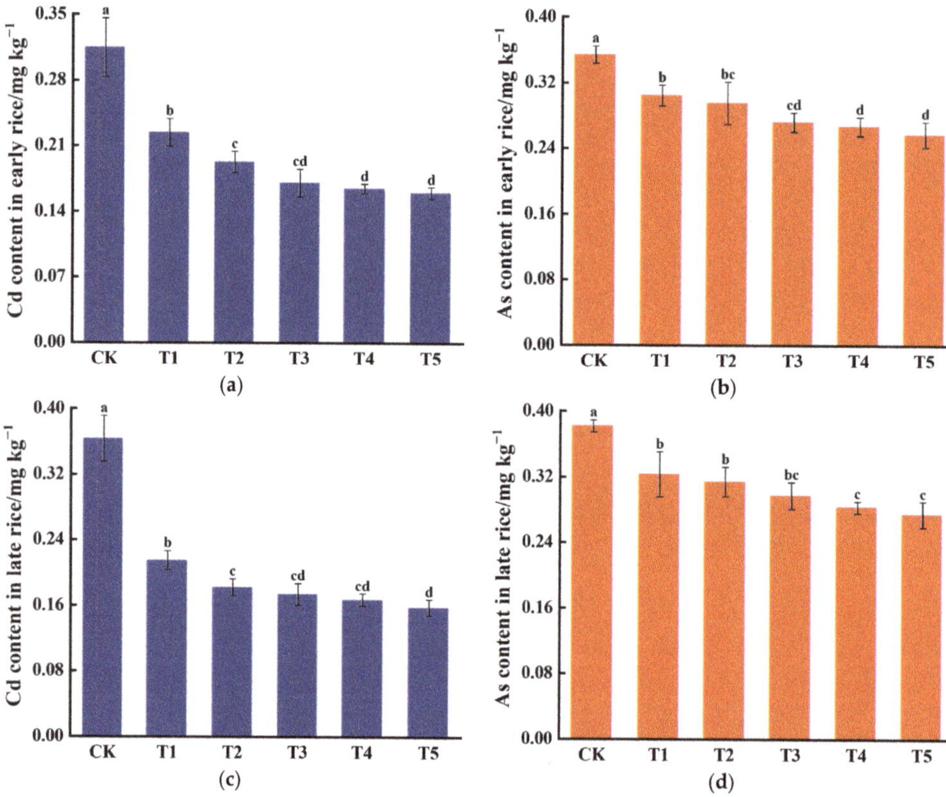

Figure 1. Effects of different treatments on Cd and As content in the grains of early rice (**a**,**b**) and late rice (**c**,**d**). CK, T1, T2, T3, T4, and T5 represent spray treatments of 0, 0.2, 0.5, 0.8, 1.2, and 1.5 mmol L^{-1} DDAC, respectively. Error bars represent SD. Different letters (a–d) indicate significant differences ($p < 0.05$) between treatments.

3.2. Effects of DDAC on the Essential Element Content of Rice Grains

The application of DDAC had a significant effect on the content of essential elements in rice grains (Figure 2). Interestingly, the contents of K, Ca, Mg, and Mn in rice varied with different rice varieties and seasons, and the contents of these elements in late rice were significantly higher than those in early rice. In both early and late rice, the contents of essential elements were in the order of K > Mg > Ca > Mn > Fe > Zn. Compared with CK, the contents of K, Ca, and Mg were significantly increased after DDAC spraying, while the contents of Fe, Mn, and Zn were significantly decreased. When 1.5 mmol L^{-1} DDAC was sprayed, the contents of K, Ca, and Mg in early rice grains increased by 9.2%, 20.1%, and 11.4%, and the contents of Fe, Mn, and Zn decreased by 12.2%, 52.1%, and 21.1%, respectively. The contents of K, Ca, and Mg in late rice grains increased by 6.9%, 12.8%, and 19.8%, respectively, and the contents of Fe, Mn, and Zn decreased by 22.5%, 27.8%, and 23.0%, respectively.

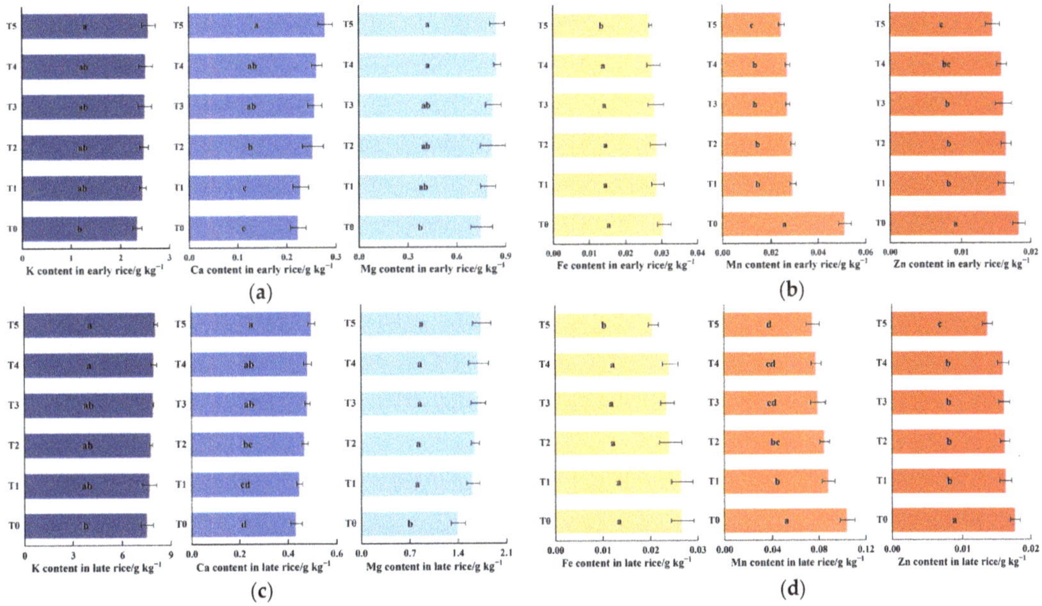

Figure 2. Effects of different treatments on essential element content in the grains of early rice (**a**,**b**) and late rice (**c**,**d**). T0, T1, T2, T3, T4, and T5 represent spray treatments of 0, 0.2, 0.5, 0.8, 1.2, and 1.5 mmol L^{-1} DDAC, respectively. Error bars represent SD. Different letters (a–d) indicate significant differences ($p < 0.05$) among treatments.

In order to further explore the relationship between exogenous DDAC, Cd, As, and the essential element content, Pearson correlation analysis was performed (Figure 3). The results showed that the spraying of DDAC had significant effects on the contents of Cd, As, K, Ca, Mg, Fe, Mn, and Zn in early and late rice grains. The content of Cd in early rice grains was negatively correlated with K (r = −0.49 *), Ca (r = −0.68 ***) and Mg (r = −0.59 **), while it was positively correlated with As (r = 0.87 ***), Fe (r = 0.60 **), Mn (r = 0.92 ***), and Zn (r = 0.73 ***). The As content was not significantly correlated with Fe, but was negatively correlated with K (r = −0.62 **), Ca (r = −0.60 **), and Mg (r = −0.52 **) and positively correlated with Mn (r = 0.86 ***) and Zn (r = 0.66 ***). The correlation between elements in late rice was similar to that in early rice. The content of Cd in late rice grains was negatively correlated with K (r = −0.57 **), Ca (r = −0.77 ***), and Mg (r = −0.73 ***), while it was positively correlated with As (r = 0.85 ***), Fe (r = 0.53 **), Mn (r = 0.83 ***), and Zn (r = 0.67 ***). The As content was negatively correlated with K (r = −0.68 ***), Ca (r = −0.71 ***), and Mg (r = −0.66 ***) and positively correlated with Fe (r = 0.58 **), Mn (r = 0.85 ***), and Zn (r = 0.74 ***).

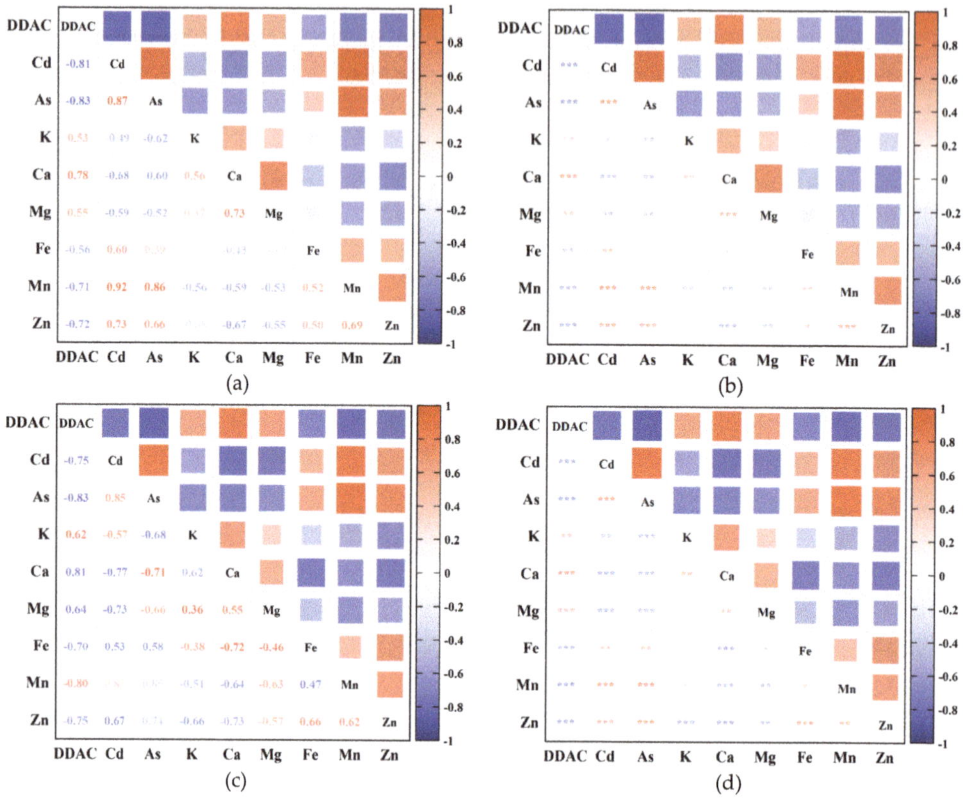

Figure 3. Person's correlation analysis of Cd, As, K, Ca, Mg, Fe, Mn, and Zn content in early rice (**a**,**b**) and late rice (**c**,**d**) upon different DDAC treatments. "*" in the panel means that the correlations were significant at $p < 0.05$, "**" means that they were significant at $p < 0.01$, and "***" means that they were significant at $p < 0.001$.

3.3. Effects of DDAC on the Total Amino Acid Content in Rice Grains

Spraying DDAC onto the leaf surface could increase the amino acid content in rice grains (Figure 4). Amino acids (AAs) are divided into essential amino acids (EAAs) and non-essential amino acids (NEAAs), which are important indexes for evaluating the nutritional quality of rice. The amino acid content of early rice and late rice also showed a certain difference; the total amount of NEAAs was higher than the total amount of EAAs. The amino acid content of early rice and late rice was also different; Glu was the most abundant amino acid in rice, and its content in early rice and late rice reached 10.32–13.16 g kg^{-1} and 9.29–12.57 g kg^{-1}, respectively. In addition, the contents of cysteine (Cys), aspartate (Asp), and valine (Val) were also relatively rich, reaching 7.72–8.81 g kg^{-1}, 7.12–7.98 g kg^{-1}, and 6.89–7.14 g kg^{-1} in early rice and 6.96–7.56 g kg^{-1}, 6.86–7.34 g kg^{-1}, and 6.46–6.63 g kg^{-1} in late rice, respectively. Methionine (Met) and histidine (His) were the two least abundant amino acids, at only 1.01–1.18 g kg^{-1} and 1.38–1.65 g kg^{-1} in early rice and 0.87–1.06 g kg^{-1} and 1.35–1.60 g kg^{-1} in late rice.

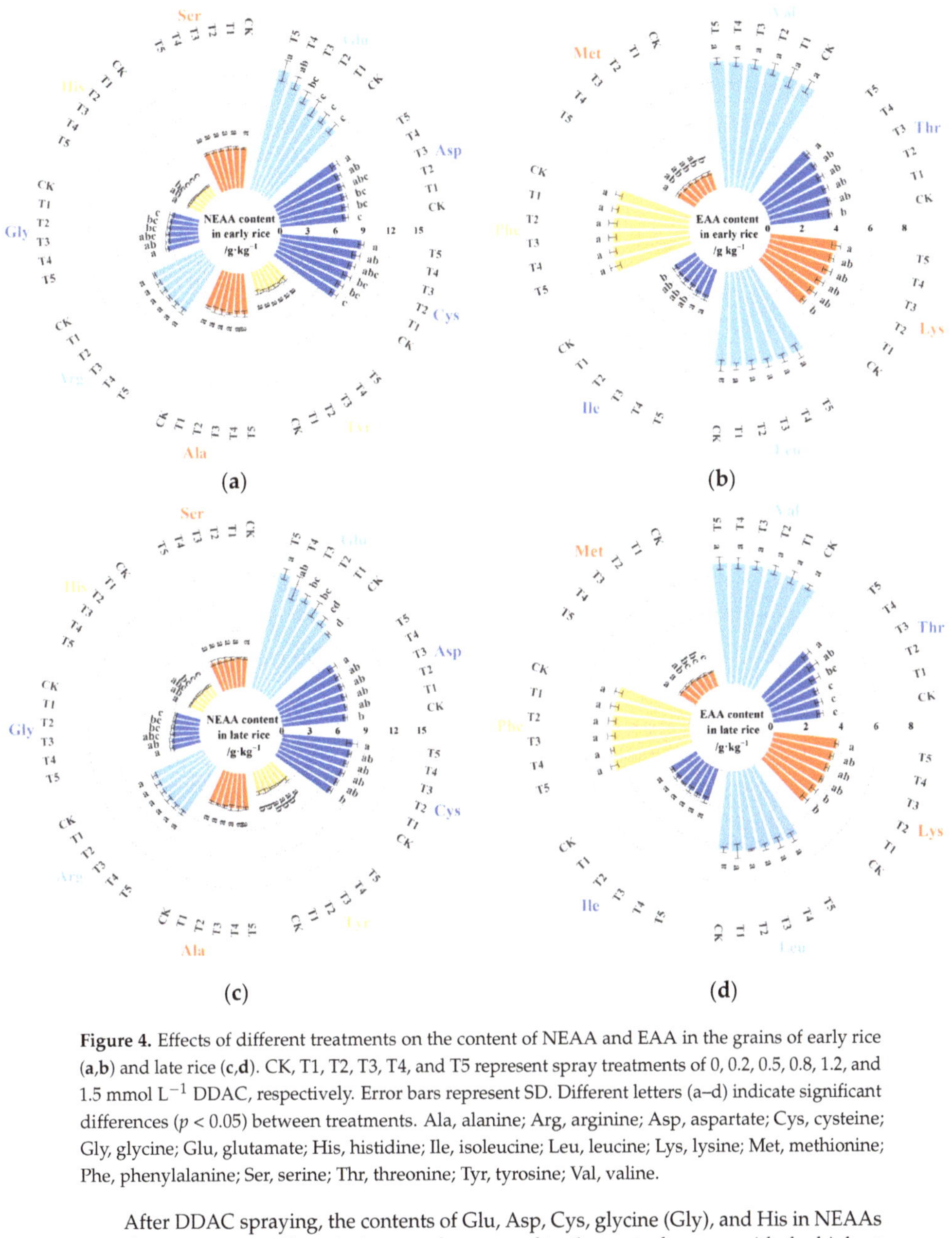

Figure 4. Effects of different treatments on the content of NEAA and EAA in the grains of early rice (**a**,**b**) and late rice (**c**,**d**). CK, T1, T2, T3, T4, and T5 represent spray treatments of 0, 0.2, 0.5, 0.8, 1.2, and 1.5 mmol L^{-1} DDAC, respectively. Error bars represent SD. Different letters (a–d) indicate significant differences ($p < 0.05$) between treatments. Ala, alanine; Arg, arginine; Asp, aspartate; Cys, cysteine; Gly, glycine; Glu, glutamate; His, histidine; Ile, isoleucine; Leu, leucine; Lys, lysine; Met, methionine; Phe, phenylalanine; Ser, serine; Thr, threonine; Tyr, tyrosine; Val, valine.

After DDAC spraying, the contents of Glu, Asp, Cys, glycine (Gly), and His in NEAAs in early rice were significantly increased compared to the control group, with the highest increases of 21.6%, 10.6%, 12.3%, 15.1%, and 16.3%, respectively. However, threonine (Thr), lysine (Lys), isoleucine (Ile), and Met in the EAAs increased significantly only when DDAC was sprayed at a higher concentration. Other amino acid contents did not change significantly. In late rice, Glu, Gly, His, Thr, and Met significantly increased after the spraying of DDAC, with the highest increases of 26.1%, 13.5%, 15.6%, 21.1%, and 17.3%, respectively. While the changes in Asp and Cys were significant under 1.50 mmol L^{-1} DDAC treatment.

The correlation analysis of Cd and As content and amino acid content showed that the accumulation of Cd and As in rice was negatively correlated with the content of amino acids (Figure 5). In particular, the Glu content was strongly correlated with both the Cd and As content. The His, Gly, Glu, Asp, Cys, Thr, Lys, Ile, and Met content were significantly correlated with Cd content in early rice. The As content was no longer significantly correlated with the Met content, but was significantly correlated with Tyr. The correlation between Cd and amino acids in late rice was similar to that in early rice. The As content and Met content also showed significant correlations in late rice.

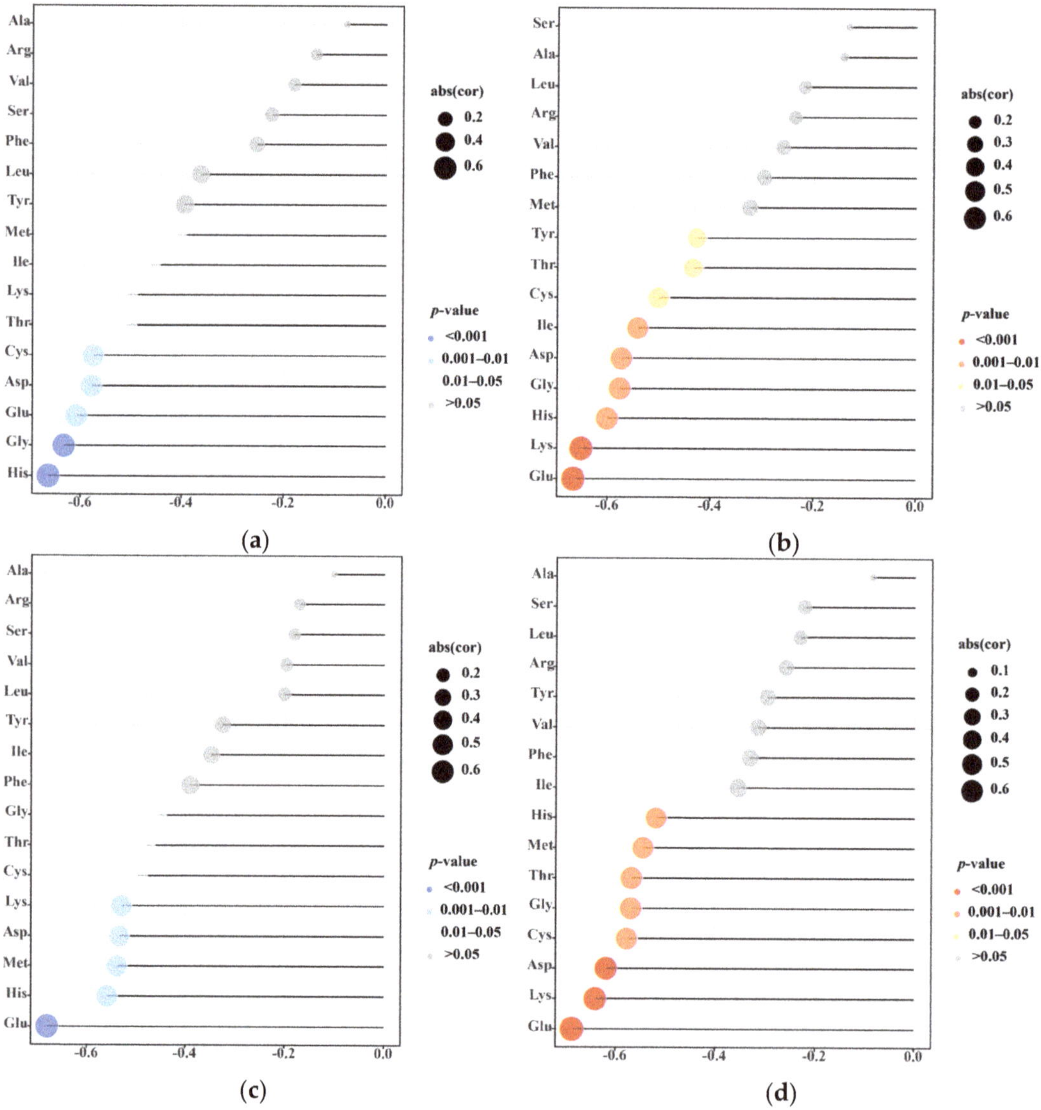

Figure 5. Pearson's correlation analysis of Cd and As content and amino acid content in early rice (**a**,**b**) and late rice (**c**,**d**). There was significant correlation between the two groups when $p < 0.05$. Abbreviations have the same meaning as in Figure 4.

4. Discussion

K, Ca, Mg, Fe, Mn, and Zn are all essential nutrients required by rice, and their absorption and transport are crucial for growth, metabolism, and development in rice [28]. The transportation of these essential cations is regulated by *OsNramp*, *OsIRT*, *OsHMA*, *OsZIP*, *OsLCT*, and other transporters, as well as non-selective cation channels (NSCCs) [29–31]. NSCCs are a collection of channel proteins located in the inner membrane of cells such as the plasma membrane and vacuole membrane, which mediate the transmembrane transport of cations with different valence states [32]. In addition, these essential cation-related transporters and NSCCs also simultaneously regulate the absorption and transport of Cd in rice. The transport of As in rice also relies on transporters of essential elements, such as P transporters (Phts), and Si transporters (*OsLsi1* and *OsLsi2*) [33–35]. While the competitive effect of essential elements on transporters and channels is greater than that of harmful elements, the transport rate of beneficial elements in plants is much higher than that of harmful elements. In this study, spraying DDAC on the leaf surface significantly decreased the Cd and As content, but significantly increased the Ca content. Due to the competitive relationship between Cd and Ca, this increase in Ca content could inhibit the absorption and transport of Cd, thereby reducing the accumulation of Cd in rice grains. Studies have shown that an increase in Ca reduces the activity of As [36]. These results indicate that DDAC application can inhibit the absorption and transport of Cd by regulating the absorption and transport of essential elements, which may be related to increasing the ability of Ca to compete with transporters and NSCCs. In addition, increases in Ca can inhibit the activity of As and reduce the transport of As.

Amino acids are related to ion transport and also play a key role in the detoxification of heavy metals [37,38]. Glu and Asp are aliphatic amino acids, which can form a ring complex with Cd through two O atoms from α-COO- and the side chain-COO-. Arginine (Arg) is a basic amino acid that can form a complex with Cd through the N and O of cyanide and carboxyl groups, thereby reducing Cd activity [39]. Glu, Gly, and Cys are precursor substances for the synthesis of glutathione (GSH, γ-Glu-Cys-Gly) and phytochelatins (PCs, γ-Glu-Cys)$_n$-Gly) [40,41]. GSH and PCs form chelates with Cd via S atoms on the sulfhydryl group, thereby reducing the activity of Cd and inhibiting the migration and transport of Cd [42]. Most plants have a strong ability to reduce As(V), and As(V) entering plant cells can be rapidly reduced to As(III) by arsenate reductase [43,44]. As(III) can form complexes with GSH and PCs, which are further isolated into the vacuole by the ABBC1/ABCC2 transporter [45]. In this study, the foliar spraying of DDAC promoted the synthesis of amino acids and alleviated amino acid metabolism disorder induced by Cd-As co-contamination. Increases in Glu, Gly, Cys, Asp, and Arg content were conducive to the synthesis of GSH. GSH, Glu, Asp, and Arg can be used as ligands to promote the chelation of heavy metals. These results indicate that the foliar spraying of DDAC increased the chelation of chelating ligands with Cd and As in rice, thereby jointly inhibiting the accumulation of Cd and As in rice grains.

In addition, the selective permeability of glutamate receptor-like channels (GLRs) is regulated by ligands and is closely related to Ca conduction and cation transport, and its selectivity for beneficial elements is much higher than that for harmful elements [46,47]. Glu, Gly, Cys, Asp, alanine (Ala), and serine (Ser) can be used as activators for GLRs to regulate the flow rate of ions entering and leaving channels [48,49]. Research has shown that the higher the activity of GLRs, the stronger the ability of rice to inhibit Cd transport, and rice with damaged GLR structures lead to excessive Cd accumulation in rice grains [50]. These results indicated that increasing GLR activity could significantly reduce the uptake and transport of Cd in rice. In this study, the leaf spraying DDAC was able to effectively increase the content of amino acids such as Glu, Asp, Ala, and Ser, which can be used as GLR activators, and may improve the recognition ability of GLRs for harmful elements, thus inhibiting the absorption and transport of Cd.

5. Conclusions

The accumulation of Cd and As in rice grains can affect the absorption of essential elements and the synthesis of amino acids. Field experiments were conducted to investigate the effects of foliar spraying of DDAC on Cd and As accumulation in rice grains in Hunan province. The results indicated that the foliar spraying of DDAC could significantly reduce the Cd and As content in rice grains. At the same time, the content of Ca in the grains was significantly increased. In addition, Glu, Gly, Cys, and other amino acid contents were significantly increased with the foliar spraying of DDAC; this was beneficial for rice, alleviating the toxicity caused by Cd and As. In conclusion, the foliar spraying of DDAC can promote amino acid synthesis and regulate the absorption and transport of essential elements in rice. Therefore, the accumulation of Cd and As in rice grains was significantly reduced. Foliar application of DDAC provides a new idea and method for solving the problem of Cd and As exceeding acceptable standards in rice.

Author Contributions: Conceptualization, L.F.; Funding acquisition, C.Z. and W.X.; Investigation, Y.D.; Methodology, L.F. and X.L.; Project administration, C.Z.; Supervision, C.Z.; Writing—original draft, L.F.; Writing—review and editing, J.D. and D.R.L. All authors have read and agreed to the published version of the manuscript.

Funding: This work was supported by the National Key Research and Development Program of China (No. 2022YFD1700103), the Foreign Experts Program of the Ministry of Science and Technology of the People's Republic of China (No. G2022051018L), and the National Natural Science Foundation of China (42207039), Funds for Science and Technology Innovation Project from the Chinese Academy of Agricultural Sciences.

Institutional Review Board Statement: Not applicable.

Informed Consent Statement: Not applicable.

Data Availability Statement: Dataset available on request from the authors.

Acknowledgments: Thanks to all the people who have helped to write and revise this article.

Conflicts of Interest: The authors declare no conflicts of interest.

References

1. Sui, F.Q.; Chang, J.D.; Tang, Z.; Liu, W.J.; Huang, X.Y.; Zhao, F.J. Nramp5 expression and functionality likely explain higher cadmium uptake in rice than in wheat and maize. *Plant Soil* **2018**, *433*, 377–389. [CrossRef]
2. Su, Y.H.; McGrath, S.P.; Zhao, F.J. Rice is more efficient in arsenite uptake and translocation than wheat and barley. *Plant Soil* **2010**, *328*, 27–34. [CrossRef]
3. Wang, Q.; Xie, Z.Y.; Li, F.B. Using ensemble models to identify and apportion heavy metal pollution sources in agricultural soils on a local scale. *Environ. Pollut.* **2015**, *206*, 227–235. [CrossRef] [PubMed]
4. Qiao, P.W.; Wang, S.; Li, J.B.; Shan, Y.; Wei, Y.; Zhang, Z.G.; Lei, M. Quantitative analysis of the contribution of sources, diffusion pathways, and receptor attributes for the spatial distribution of soil heavy metals and their nested structure analysis in China. *Sci. Total Environ.* **2023**, *882*, 163647. [CrossRef] [PubMed]
5. Wei, R.H.; Chen, C.; Kou, M.; Liu, Z.Y.; Wang, Z.; Cai, J.X.; Tan, W.F. Heavy metal concentrations in rice that meet safety standards can still pose a risk to human health. *Commun. Earth Environ.* **2023**, *4*, 84. [CrossRef]
6. Zhao, F.J.; Wang, P. Arsenic and cadmium accumulation in rice and mitigation strategies. *Plant Soil* **2020**, *446*, 1–21. [CrossRef]
7. Honma, T.; Ohba, H.; Kaneko-Kadokura, A.; Makino, T.; Nakamura, K.; Katou, H. Optimal Soil Eh, pH, and Water Management for Simultaneously Minimizing Arsenic and Cadmium Concentrations in Rice Grains. *Environ. Sci. Technol.* **2016**, *50*, 4178–4185. [CrossRef] [PubMed]
8. Zeng, P.; Wei, B.Y.; Zhou, H.; Gu, J.F.; Liao, B.H. Co-application of water management and foliar spraying silicon to reduce cadmium and arsenic uptake in rice: A two-year field experiment. *Sci. Total Environ.* **2022**, *818*, 151801. [CrossRef] [PubMed]
9. Gu, J.F.; Zhou, H.; Tang, H.L.; Yang, W.T.; Zeng, M.; Liu, Z.M.; Peng, P.Q.; Liao, B.H. Cadmium and arsenic accumulation during the rice growth period under in situ remediation. *Ecotoxicol. Environ. Saf.* **2019**, *171*, 451–459. [CrossRef]
10. Xue, W.J.; Zhang, C.B.; Wang, P.P.; Wang, C.R.; Huang, Y.C.; Zhang, X.; Liu, Z.Q. Rice vegetative organs alleviate cadmium toxicity by altering the chemical forms of cadmium and increasing the ratio of calcium to manganese. *Ecotoxicol. Environ. Saf.* **2019**, *184*, 109640. [CrossRef]

11. Moulick, D.; Ghosh, D.; Mandal, J.; Bhowmick, S.; Mondal, D.; Choudhury, S.; Santra, S.C.; Vithanage, M.; Biswas, J.K. A cumulative assessment of plant growth stages and selenium supplementation on arsenic and micronutrients accumulation in rice grains. *J. Clean. Prod.* **2023**, *386*, 135764. [CrossRef]
12. Cui, H.; Tang, S.T.; Huang, S.Q.; Lei, L.D.; Jiang, Z.M.; Li, L.; Wei, S.Q. Simultaneous mitigation of arsenic and cadmium accumulation in rice grains by foliar inhibitor with ZIF-8@Ge-132. *Sci. Total Environ.* **2023**, *860*, 160307. [CrossRef] [PubMed]
13. Yao, A.J.; Yang, J.L.; Liu, Y.; Su, G.Q.; Zhao, M.; Wang, S.Z.; Tang, Y.T.; Qiu, R.L. Mitigation effects of foliar supply of different sulfur forms on uptake, translocation and grain accumulation of Cd and As by paddy rice on basis of liming. *Sci. Total Environ.* **2023**, *905*, 167338. [CrossRef] [PubMed]
14. Li, H.; Luo, N.; Li, Y.W.; Cai, Q.Y.; Li, H.Y.; Mo, C.H.; Wong, M.H. Cadmium in rice: Transport mechanisms, influencing factors, and minimizing measures. *Environ. Pollut.* **2017**, *224*, 622–630. [CrossRef] [PubMed]
15. Zhao, P.W.; Huang, P.C.; Yan, X.; Chukwuma, A.; Yang, S.; Yang, Z.H.; Li, H.; Yang, W.C. Inhibitory effect of exogenous mineral elements (Si, P, Zn, Ca, Mn, Se, Fe, S) on rice Cd accumulation and soil Cd bioavailability in Cd-contaminated farmlands: A meta-analysis. *Chemosphere* **2023**, *343*, 140282. [CrossRef] [PubMed]
16. Abbas, G.; Murtaza, B.; Bibi, I.; Shahid, M.; Niazi, N.K.; Khan, M.I.; Amjad, M.; Hussain, M.; Natasha. Arsenic Uptake, Toxicity, Detoxification, and Speciation in Plants: Physiological, Biochemical, and Molecular Aspects. *Int. J. Environ. Res. Public Health* **2018**, *15*, 59. [CrossRef]
17. Kumar, S.; Kumar, S.; Mohapatra, T. Interaction between macro- and micro- nutrients in plants. *Front. Plant Sci.* **2021**, *12*, 665583. [CrossRef] [PubMed]
18. Yang, J.C.; Zhou, Y.J.; Jiang, Y. Amino acids in rice grains and their regulation by polyamines and phytohormones. *Plants* **2022**, *11*, 1581. [CrossRef] [PubMed]
19. Galili, G.; Amir, R.; Fernie, A.R. The regulation of essential amino acid synthesis and accumulation in plants. *Annu. Rev. Plant Biol.* **2016**, *67*, 153–178. [CrossRef]
20. Chen, J.; Le, X.C.; Zhu, L.Z. Metabolomics and transcriptomics reveal defense mechanism of rice (*Oryza sativa*) grains under stress of 2,2′,4,4′-tetrabromodiphenyl ether. *Environ. Int.* **2019**, *133*, 105154. [CrossRef]
21. Jiang, M.; Jiang, J.; Li, S.; Li, M.; Tan, Y.Y.; Song, S.Y.; Shu, Q.Y.; Huang, J.Z. Glutamate alleviates cadmium toxicity in rice via suppressing cadmium uptake and translocation. *J. Hazard. Mater.* **2020**, *384*, 121319. [CrossRef] [PubMed]
22. Asgher, M.; Sehar, Z.; Rehaman, A.; Rashid, S.; Ahmed, S.; Per, T.S.; Alyemeni, M.N.; Khan, N.A. Exogenously-applied L-glutamic acid protects photosynthetic functions and enhances arsenic tolerance through increased nitrogen assimilation and antioxidant capacity in rice (*Oryza sativa* L.). *Environ. Pollut.* **2022**, *301*, 119008. [CrossRef] [PubMed]
23. Liu, S.Y.; Fu, L.; Zhang, C.B.; Deng, J.W.; Xue, W.J.; Deng, Y. Effects of exogenous chlorinated amino acetic acid on cadmium and mineral elements in rice seedlings. *Toxics* **2023**, *11*, 71. [CrossRef] [PubMed]
24. Fu, L.; Deng, J.W.; Liu, S.Y.; Zhang, C.B.; Xue, W.J.; Mailhot, G.; Vione, D.; Deng, Y.; Wang, C.R.; Wang, L. Efficient regulation of cadmium accumulation by carboxymethylammonium chloride in rice: Correlation analysis and expression of transporter gene *OsGLR3*. *Sci. Total Environ.* **2024**, *930*, 172861. [CrossRef] [PubMed]
25. Li, Y.; Chu, Y.T.; Sun, H.Y.; Bao, Q.L.; Huang, Y.Z. Melatonin alleviates arsenite toxicity by decreasing the arsenic accumulation in cell protoplasts and increasing the antioxidant capacity in rice. *Chemosphere* **2023**, *312*, 137292. [CrossRef] [PubMed]
26. Yang, X.R.; Wang, C.R.; Huang, Y.C.; Liu, B.; Liu, Z.Q.; Huang, Y.Z.; Cheng, L.L.; Huang, Y.F.; Zhang, C.B. Foliar application of the sulfhydryl compound 2,3-dimercaptosuccinic acid inhibits cadmium, lead, and arsenic accumulation in rice grains by promoting heavy metal immobilization in flag leaves. *Environ. Pollut.* **2021**, *285*, 117355. [CrossRef] [PubMed]
27. Yuan, K.; Wang, C.R.; Zhang, C.B.; Huang, Y.C.; Wang, P.P.; Liu, Z.Q. Rice grains alleviate cadmium toxicity by expending glutamate and increasing manganese in the cadmium contaminated farmland. *Environ. Pollut.* **2020**, *262*, 114236. [CrossRef]
28. Kumarathilaka, P.; Bundschuh, J.; Seneweera, S.; Ok, Y.S. An integrated approach of rice hull biochar-alternative water management as a promising tool to decrease inorganic arsenic levels and to sustain essential element contents in rice. *J. Hazard. Mater.* **2021**, *405*, 124188. [CrossRef] [PubMed]
29. Zou, M.M.; Zhou, S.L.; Zhou, Y.J.; Jia, Z.Y.; Guo, T.W.; Wang, J.X. Cadmium pollution of soil-rice ecosystems in rice cultivation dominated regions in China: A review. *Environ. Pollut.* **2021**, *280*, 116965. [CrossRef]
30. Feng, X.M.; Han, L.; Chao, D.Y.; Liu, Y.; Zhang, Y.J.; Wang, R.G.; Guo, J.K.; Feng, R.W.; Xu, Y.M.; Ding, Y.Z.; et al. Ionomic and transcriptomic analysis provides new insight into the distribution and transport of cadmium and arsenic in rice. *J. Hazard. Mater.* **2017**, *331*, 246–256. [CrossRef]
31. Tan, L.T.; Qu, M.M.; Zhu, Y.X.; Peng, C.; Wang, J.R.; Gao, D.Y.; Chen, C.Y. ZINC TRANSPORTER5 and ZINC TRANSPORTER9 function synergistically in zinc/cadmium uptake. *Plant Physiol.* **2020**, *183*, 1235–1249. [CrossRef] [PubMed]
32. Han, X.X.; Zhang, C.B.; Wang, C.R.; Huang, Y.C.; Liu, Z.Q. Gadolinium inhibits cadmium transport by blocking non-selective cation channels in rice seedlings. *Ecotoxicol. Environ. Saf.* **2019**, *179*, 160–166. [CrossRef]
33. Cao, Y.; Sun, D.; Ai, H.; Mei, H.Y.; Liu, X.; Sun, S.B.; Xu, G.H.; Liu, Y.G.; Chen, Y.S.; Ma, L.N.Q. Knocking out *OsPT4* gene decreases arsenate uptake by rice plants and inorganic arsenic accumulation in rice grains. *Environ. Sci. Technol.* **2017**, *51*, 12131–12138. [CrossRef]
34. Shi, G.; Ma, H.; Chen, Y.; Liu, H.; Song, G.; Cai, Q.; Lou, L.; Rengel, Z. Low arsenate influx rate and high phosphorus concentration in wheat (*Triticum aestivum* L.): A mechanism for arsenate tolerance in wheat plants. *Chemosphere* **2019**, *214*, 94–102. [CrossRef] [PubMed]

35. Kumar, A.; Basu, S.; Rishu, A.K.; Kumar, G. Revisiting the mechanisms of arsenic uptake, transport and detoxification in plants. *Environ. Exp. Bot.* **2022**, *194*, 104730. [CrossRef]
36. Zeng, F.R.; Nazir, M.M.; Ahmed, T.; Noman, M.; Ali, S.; Rizwan, M.; Alam, M.S.; Lwalaba, J.L.W.; Zhang, G.P. Calcium and L-glutamate present the opposite role in managing arsenic in barley. *Environ. Pollut.* **2023**, *321*, 121141. [CrossRef] [PubMed]
37. Kaur, G.; Asthir, B. Proline: A key player in plant abiotic stress tolerance. *Biol. Plant.* **2015**, *59*, 609–619. [CrossRef]
38. Yang, Q.Q.; Zhao, D.S.; Liu, Q.Q. Connections between amino acid metabolisms in plants: Lysine as an example. *Front. Plant Sci.* **2020**, *11*, 928. [CrossRef]
39. Xue, W.J.; Zhang, C.B.; Huang, Y.C.; Wang, C.R.; Zhang, X.; Liu, Z.Q. Rice organs concentrate cadmium by chelation of amino acids containing dicarboxyl groups and enhance risks to human and environmental health in Cd-contaminated areas. *J. Hazard. Mater.* **2022**, *426*, 128130. [CrossRef]
40. Cao, Z.Z.; Qin, M.L.; Lin, X.Y.; Zhu, Z.W.; Chen, M.X. Sulfur supply reduces cadmium uptake and translocation in rice grains (*Oryza sativa L.*) by enhancing iron plaque formation, cadmium chelation and vacuolar sequestration. *Environ. Pollut.* **2018**, *238*, 76–84. [CrossRef]
41. Zhang, J.; Hamza, A.; Xie, Z.M.; Hussain, S.; Brestic, M.; Tahir, M.A.; Ulhassan, Z.; Yu, M.; Allakhverdiev, S.I.; Shabala, S. Arsenic transport and interaction with plant metabolism: Clues for improving agricultural productivity and food safety. *Environ. Pollut.* **2021**, *290*, 117987. [CrossRef] [PubMed]
42. Xue, W.J.; Zhang, X.; Zhang, C.B.; Wang, C.R.; Huang, Y.C.; Liu, Z.Q. Mitigating the toxicity of reactive oxygen species induced by cadmium via restoring citrate valve and improving the stability of enzyme structure in rice. *Chemosphere* **2023**, *327*, 138511. [CrossRef]
43. Shi, S.L.; Wang, T.; Chen, Z.; Tang, Z.; Wu, Z.C.; Salt, D.E.; Chao, D.Y.; Zhao, F.J. *OsHAC1;1* and *OsHAC1;2* function as arsenate reductases and regulate arsenic accumulation. *Plant Physiol.* **2016**, *172*, 1708–1719. [CrossRef]
44. Xu, J.M.; Shi, S.L.; Wang, L.; Tang, Z.; Lv, T.T.; Zhu, X.L.; Ding, X.M.; Wang, Y.F.; Zhao, F.J.; Wu, Z.C. *OsHAC4* is critical for arsenate tolerance and regulates arsenic accumulation in rice. *New Phytol.* **2017**, *215*, 1090–1101. [CrossRef] [PubMed]
45. Song, W.Y.; Park, J.; Mendoza-Cózatl, D.G.; Suter-Grotemeyer, M.; Shim, D.; Hörtensteiner, S.; Geisler, M.; Weder, B.; Rea, P.A.; Rentsch, D.; et al. Arsenic tolerance in Arabidopsis is mediated by two ABCC-type phytochelatin transporters. *Proc. Natl. Acad. Sci. USA* **2010**, *107*, 21187–21192. [CrossRef]
46. Shao, Q.L.; Gao, Q.F.; Lhamo, D.; Zhang, H.S.; Luan, S. Two glutamate- and pH-regulated Ca^{2+} channels are required for systemic wound signaling in Arabidopsis. *Sci. Signal.* **2020**, *13*, eaba1453. [CrossRef]
47. Kong, D.; Hu, H.C.; Okuma, E.; Lee, Y.; Lee, H.S.; Munemasa, S.; Cho, D.; Ju, C.; Pedoeim, L.; Rodriguez, B.; et al. L-Met Activates Arabidopsis GLR Ca^{2+} Channels Upstream of ROS Production and Regulates Stomatal Movement. *Cell Rep.* **2016**, *17*, 2553–2561. [CrossRef]
48. Vatsa, P.; Chiltz, A.; Bourque, S.; Wendehenne, D.; Garcia-Brugger, A.; Pugin, A. Involvement of putative glutamate receptors in plant defence signaling and NO production. *Biochimie* **2011**, *93*, 2095–2101. [CrossRef] [PubMed]
49. Qi, Z.; Stephens, N.R.; Spalding, E.P. Calcium entry mediated by GLR3. 3, an arabidopsis glutamate receptor with a broad agonist profile. *Plant Physiol.* **2006**, *142*, 963–971. [CrossRef]
50. Zhang, X.; Xue, W.J.; Zhang, C.B.; Wang, C.R.; Huang, Y.C.; Wang, Y.T.; Peng, L.C.; Liu, Z.Q. Cadmium pollution leads to selectivity loss of glutamate receptor channels for permeation of $Ca^{2+}/Mn^{2+}/Fe^{2+}/Zn^{2+}$ over Cd^{2+} in rice plant. *J. Hazard. Mater.* **2023**, *452*, 131342. [CrossRef]

toxics

MDPI

Article

Citric Acid Inhibits Cd Absorption and Transportation by Improving the Antagonism of Essential Elements in Rice Organs

Kexin Chen [1,†], Bozhen Yu [1,†], Weijie Xue [1,*], Yuebing Sun [1], Changbo Zhang [1], Xusheng Gao [1], Xiaojia Zhou [1], Yun Deng [2], Jiarun Yang [3] and Boqian Zhang [3]

[1] Key Laboratory of Original Agro-Environmental Pollution Prevention and Control, Agro-Environmental Protection Institute, Ministry of Agriculture and Rural Affairs, Tianjin 300191, China; ckx_1_1@163.com (K.C.); 18604519599@163.com (B.Y.); sunyuebing@caas.cn (Y.S.); zhangchangbo@caas.cn (C.Z.); 13001303720@163.com (X.G.); zxj103698@163.com (X.Z.)
[2] School of Environment and Ecology, Jiangnan University, Wuxi 214122, China; dengyun@jiangnan.edu.cn
[3] College of Water Conservancy Engineering, Tianjin Agricultural University, Tianjin 300392, China; 18987184564@163.com (J.Y.); 15702675200@163.com (B.Z.)
* Correspondence: tzfelicity@163.com
† These authors contributed equally to this work.

Abstract: Excessive cadmium (Cd) in rice is a global environmental problem. Therefore, reducing Cd content in rice is of great significance for ensuring food security and human health. A field experiment was conducted to study the effects of foliar application of citric acid (CA) on Cd absorption and transportation in rice under high Cd-contaminated soils (2.04 mg·kg^{-1}). This study revealed that there was a negative correlation between Cd content in vegetative organs and CA content, and that foliar spraying of CA (1 mM and 5 mM) significantly increased CA content and reduced Cd content in vegetative organs. The Cd reduction effect of 5 mM CA was better than that of 1 mM, and 5 mM CA reduced Cd content in grains and spikes by 52% and 37%, respectively. CA significantly increased Mn content in vegetative organs and increased Ca/Mn ratios in spikes, flag leaves, and roots. CA significantly reduced soluble Cd content in vegetative organs and promoted the transformation of Cd into insoluble Cd, thus inhibiting the transport of Cd from vegetative organs to grains. The foliar field application of 1 mM and 5 mM CA could inhibit Cd absorption and transportation by reducing Cd bioactivity and increasing the antagonistic of essential elements in rice vegetative organs. These results provide technical support and a theoretical basis for solving the problem of excessive Cd in rice.

Keywords: Cd; rice; citric acid; foliar application; essential element

1. Introduction

Due to industrial development and human activities, soil cadmium (Cd) pollution has become increasingly severe (reaching almost 7.75% on Chinese farmland) [1]. As the awareness of food safety grows, preventing the accumulation of Cd in cereal crops and its transmission through the food chain has become an urgent issue [2]. As one of the primary cereal crops, rice is at an increased risk of Cd exposure in Cd-contaminated soil. The excessive accumulation of Cd in rice not only affects the normal growth and development of rice quality and yield but also poses a threat to human health through the food chain, such as causing itai-itai disease and cancer [3]. Therefore, controlling rice Cd pollution has become a hot topic in recent research. With a deeper understanding of Cd absorption and transport mechanisms in rice, various approaches are widely applied to reduce rice Cd content, including soil improvement, bioremediation techniques, and molecular biology methods [4,5]. Among them, the foliar application of conditioning agents is an economically effective mean to inhibit Cd absorption and transport [6,7].

Citric acid (CA) is a crucial substance in many core biological metabolic reactions. As a small-molecule organic acid, it has been confirmed to enhance plant tolerance to heavy metals, maintain cellular metabolism, and improve resistance to heavy metal stress [8,9]. Research has shown that citric acid-modified biochar can reduce the bioavailability of Cd in soil [10]. Citric acid could regulate the balance of anions and cations in rice by mediating the balance of cations and anions within plants and subsequently affecting the transport of mineral elements within plant tissues [11,12]. Exposed to 50 μM Cr stress, the CA content significantly increased in rice. Under Al stress, the roots of rice could enhance CA content to counter heavy metal toxicity [13,14]. Furthermore, the external application of CA has been found to enhance antioxidant capacity in pea seedlings, thereby increasing their tolerance to Cu [15]. Additionally, CA upregulated the iron transport protein (*OsIRT1*), competed with Cd-related transport proteins, and thus mitigated Cd stress in rice [16,17]. However, research has also indicated that CA enhanced the hyperaccumulation of heavy metal ions in plants [18,19]. Our earlier studies revealed that Cd stress induced rice roots and grains to secrete more CA to mitigate Cd toxicity. Different organs of rice could regulate the transport of Cd into grains. When the stress of Cd was severe, the intrinsic protective mechanisms of these organs were gradually weakened in resisting Cd stress [20]. It remains unclear whether the foliar application of CA could further enhance the blocking effect of Cd in various organs of rice to reduce the accumulation of Cd in rice under high Cd stress conditions.

The transport and distribution of mineral elements after organic acid application is also one way for plants to resist Cd stress. As basic crop nutrients, mineral elements (K, Ca, Mg, Mn, Fe, Zn, etc.) could affect the ability to absorb and accumulate Cd in plants [21]. For example, Mn, Fe, and Cd were absorbed and transported in the same pathway, thus the reduction in Mn and Fe content in rice promoted Cd accumulation in shoots [22]. The deficiency of Ca and Mg enhanced the toxicity of Cd in roots and shoots of rice seedlings by decreasing the antioxidant capacity of rice. The transport of Cu caused oxidative stress, triggering the signaling pathways in *Arabidopsis thaliana*, thus enhancing Cd resistance [23–25]. It has also been found that Cd could enhance K translocation from hyacinth roots to other tissues and reduce the uptake of Zn [26], while exogenous CA also influenced the absorption capacity of heavy metals by regulating plant mineral element content. There are no specific transport channels for Cd in plants, it mainly relies on the transport channels of other mineral elements, such as Mn transporters, so there is a competitive effect between Cd and Mn. It has now been shown that, under Cd stress, CA could modulate Mn transport to inhibit the uptake of Cd [27]. Treatment with 50 μM Mn increased the content of CA by 55.3% in the plant, thereby reducing Cd-induced oxidative damage [28]. However, the effect of the foliar application of CA on the uptake and transport of mineral elements in various organs of rice is still uncertain.

The period from panicle initiation to flowering is the main phase for rice to absorb chemical elements [29]. The flowering stage is the key period for Cd to be reactivated, redistributed, and transported to the grain within different organs of the rice, ultimately directing it toward the grains [20,30]. Spraying foliar regulators during flowering could inhibit the remigration of Cd in nutrient organs, avoiding excessive transport of Cd from vegetative organs to grains. Spraying 5 mM nano-silicon (Si) 1–2 times from the tillering to the grouting stage of rice could reduce Cd content by 31.6% and 36.1% in rice and rachises, respectively [31]. It has been found that the flowering stage is also an important period to increase grain yield and reduce the influence of Cd [32–34]. In addition, citric acid can also chelate with Cd, reducing Cd activity in vegetative organs and thus avoiding Cd transportation to rice grains.

Therefore, field experiments were conducted to explore the effects of foliar spraying of CA on the absorption and transport of Cd and essential elements in different organs of rice under high Cd stress. By analyzing the correlation between CA and Cd in different organs of rice and the differences in Cd transfer factors, the mechanism of foliar spraying of CA to reduce the content of Cd in rice was explored, which may provide a practical basis for subsequent research to develop novel foliar conditioning agents for the control of Cd pollution in rice.

2. Materials and Methods

2.1. Plant Materials and Experimental Site

A high-yielding and high-quality late modern indica rice "Huazhan" (widely cultivated in Hunan Province) was used in this study. The cultivation took place in red soil paddy fields located in the northeastern part of Hunan Province (N: 28°49′, E: 112°50′). This region experiences relatively high Cd contamination in rice fields, with a Cd content of 2.04 $mg \cdot kg^{-1}$ and similar fundamental chemical properties. Specifically, topsoil pH was 5.48; CEC was 20.2 $cmol \cdot kg^{-1}$; organic matter content was 43.85 $g \cdot kg^{-1}$; K, Ca, Mg, and Fe were 10.1, 21.2, 30.8, and 32.9 $g \cdot kg^{-1}$, respectively; Mn and Zn were 286.4 and 125.8 $mg \cdot kg^{-1}$.

2.2. Experimental Design and Treatments

The experiment included a control group (CK, treated with surfactant) and two treatment groups (S1: treated with 1 mM CA + surfactant; S2: treated with 5 mM CA + surfactant). We employed a completely randomized block design, with three replicates for each group. The plot dimensions were 2.5 m (length) × 4 m (width), resulting in an area of 10 m^2. Rice was sown in June 2018 and transplanted to the paddy fields in late July. To avoid any impact on the soil, during the rice flowering stage (late September), different concentrations of citric acid were uniformly sprayed on rice leaves. Field management practices were consistent with those used in high-yield rice fields in Hunan.

2.3. Sample Processed and Collection

During the rice maturation stage, fresh rice samples were collected from each paddy field using a five-point sampling method. After washing with deionized water, the samples were divided into the following components: grains, panicles (including the sheath), flag leaves, spikes, necks, remaining shoot parts, and roots. Each component was further split into two portions.

One portion was dried at 105 °C in an oven for 30 min, followed by drying at 70 °C to constant weight. The dried samples were ground through a 20-mesh nylon sieve and used for Cd and essential element analysis after digestion. The other portion was air-dried naturally in a cool ventilated area and used to determine different forms of Cd content in the plant organs.

2.4. Sample Parameter Determination

2.4.1. Cd and CA Content Determination

Approximately 0.2500 g of dried rice samples from different plant parts was weighed into digestion tubes. Nitric acid (65.0–68.0 wt%, 8 mL, analytical reagent) was added, and the samples were soaked for 12 h. Subsequently, the samples were digested in an electric digestion instrument at 80 °C for 1.5 h, followed by 120 °C for 1.5 h, and then heated to 150 °C for 3 h. After digestion, the samples were heated to 160–180 °C to drive off excess acid (leaving 1 mL of digestion solution in the tube). The solution was cooled, diluted to 50 mL in a volumetric flask with distilled water, filtered, and analyzed using inductively coupled plasma mass spectrometry (ICP-MS, Agilent 7500a, Santa Clara, CA, USA).

For CA content determination, a plant citric acid assay kit (Solaibao, Beijing, China) was used for extraction and pretreatment. Absorbance was measured using a UV spectrophotometer, and the CA content was calculated based on a standard curve.

2.4.2. Extraction of Cd Speciation

We weighed 0.3 g of plant samples and added 30 mL of 80% ethanol (sample-to-extractant ratio w:v = 1:100). The mixture was homogenized and shaken at 25 °C for 22 h. Then, we centrifuged it at 5000× g for 10 min, and the supernatant was collected in a clean container. We repeated this process twice and combined the supernatants from the three centrifugations. Next, we sequentially extracted the samples using different extractants. For each extractant, we evaporated the collected solution and the sample residue was dried on an electroplated plate at 70 °C to constant weight for subsequent analysis. The extraction and quantification of Cd chemical forms were adjusted based on previous methods [35,36]. The different extractants used were as follows: (1) 80% ethanol extraction (FE-Cd): extracts inorganic Cd; (2) deionized water extraction (FW-Cd): extracts water-soluble Cd. The two collected supernatants were combined to obtain soluble Cd, and the residual part was collected to obtain insoluble Cd.

2.4.3. Determination of Mineral Element Content

An inductively coupled plasma mass spectrometer (ICP-MS, Agilent 7500a, Santa Clara, CA, USA) was used to measure the content of K, Mg, Ca, Fe, Mn, and Zn. The working conditions for ICP-MS were as follows: radiofrequency power of 1350W, plasma flow rate of 15.0 $L \cdot min^{-1}$, carrier gas flow rate of 1.12 $L \cdot min^{-1}$, and peristaltic pump speed of 6.0 rmp. A mixed standard solution was prepared (Agilent, Part # 5183-4688), with concentrations of 1000 $\mu g \cdot mL^{-1}$ for Ca, Fe, and K, and 10 $\mu g \cdot mL^{-1}$ for Mn and Zn. For the preparation of on-machine standard solutions, the concentration ranges were as follows: K, Ca, and Fe ranged from 0 to 40,000 $\mu g \cdot L^{-1}$, while Mn and Zn ranged from 0 to 400 $\mu g \cdot L^{-1}$. Standard curves were generated for each element.

2.4.4. Transfer Factor Calculation

The transfer factor (TF) was used to analyze the transfer ability of soluble Cd between different rice organs. The calculation followed a method used by previous researchers [37]. The specific formula is as follows:

$$TF_{A/B} = \frac{T_A}{S_B} \quad (1)$$

where $TF_{A/B}$ stands for the transfer factor of soluble Cd between adjacent organs, T_A represents total Cd content in adjacent upper organs (mg/kg), and S_B stands for soluble Cd content in adjacent lower organs (mg/kg).

2.5. Data Analysis

The experimental data represent the average of three replicates. Significance analysis was performed using IBM SPSS Statistics 26, employing ANOVA and LSD methods to compare differences between different treatments ($p < 0.05$). Microsoft Excel 2021 and Origin 2022 were used for table and chart plotting.

3. Results

3.1. Effects of Leaf Spraying with CA on CA and Cd in Rice Organs

3.1.1. CA and Cd Content and Correlation

The ability of Cd accumulation in rice organs varied significantly. The Cd content in rice roots was 7.53 $mg \cdot kg^{-1}$ under only Cd-contaminated treatment, while, among the shoots, the average Cd concentration was found in the spike nodes, reaching 30.33 $mg \cdot kg^{-1}$. The inverted nodes had a Cd content as high as 17.59 $mg \cdot kg^{-1}$, which was 4.03 and 2.34 times higher than in the roots, respectively. However, the Cd content in the rice grains was relatively low, approximately 0.90 $mg \cdot kg^{-1}$, exceeding both China's grain safety standard (0.25 $mg \cdot kg^{-1}$) and the international grain safety standard (0.4 $mg \cdot kg^{-1}$). During the rice flowering stage, the foliar application of CA significantly increased the CA content in grains and vegetative organs, while reducing the Cd content in rice grains and their associated organs. The effect of spraying 5 mM CA was superior to that of 1 mM CA. Specifically, 1 mM and 5 mM CA significantly reduced Cd content in grains by 27.54% and 52.28%, respectively. Additionally, 5 mM CA significantly decreased Cd content in the necks of spikes, spike nodes, flag leaves, inverted second internodes, inverted nodes, and the inverted second leaves (Figure 1a). The lowest CA content was observed in rice grains and roots, while the CA levels in spike nodes were approximately 37.22 and 30.67 times higher than those in grains and roots, respectively. Furthermore, high concentrations of CA significantly increased the CA content in the spike nodes, flag leaves, inverted nodes, inverted second leaves, and roots, with no change in the necks of spikes, inverted second internodes, and old organizations (Figure 1b).

(a)

(b)

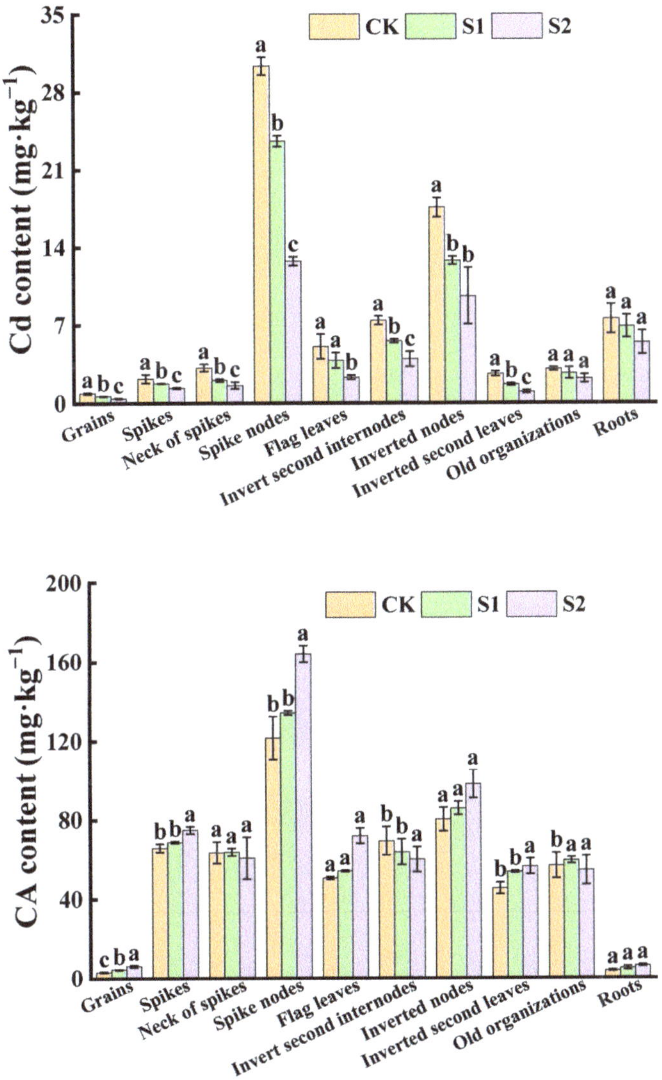

Figure 1. Content of Cd (**a**) and CA (**b**) in different tissues of rice. Different letters indicate significant difference ($p < 0.05$) among treatments in the same organ according to Duncan's test.

The Cd content of rice grains was positively correlated with the Cd content of shoot organs but not significantly correlated with the Cd content of the root system (Table 1). Except for rachis Cd content and old tissue Cd content, Cd content was positively correlated with shoot organs of rice. Citric acid content was significantly correlated among all rice organs except spike necks, binodal internodes, and old tissues. Cd and citric acid contents in seeds, rachis, nodes, flag leaves, inverted nodes, inverted leaves, and roots were negatively correlated, with the highest correlation coefficients of −0.92 and −0.94 between Cd and citric acid contents in seeds and nodes.

Table 1. Correlation between Cd and CA content in rice organisms.

Organ of Rice	Grain	Spike	Neck of Spike	Spike Knots	Flag Leaf	Inverted Two Internodes	Inverted Node	Inverted Two Leaves	Others	Root
Grains		0.90 **	——	0.94 **	0.96 **	——	0.85 **	0.84 **	——	0.93 **
Spikes	0.89 **		——	0.93 **	0.89 **	——	0.80 *	0.81 **	——	0.88 **
Neck of spikes	0.84 **	0.79 *		——	——	——	——	——	0.88 **	——
Spike nodes	0.97 **	0.88 **	0.90 **		0.95 **	——	0.80 **	0.82 **	——	0.92 **
Flag leaves	0.77 *	0.67 *	0.92 **	0.85 **		——	0.84 **	0.75 *	——	0.88 **
Inverted second internodes	0.98 **	0.85 **	0.87 **	0.99 **	0.81 **		——	——	——	——
Inverted nodes	0.95 **	0.77 *	0.78 *	0.93 **	0.67 *	0.96 **		——	——	0.88 **
Inverted second leaves	0.93 **	0.88 **	0.95 **	0.96 **	0.91 **	0.95 **	0.84 **		——	0.79 *
Others	0.78 *	——	0.71 *	0.85 **	0.58	0.87 **	0.89 **	0.76 *		——
Roots	——	0.70 *	——	0.72 *	0.77 *	0.71 *	——	0.81 **	——	
Cd/CA in the same organ	−0.92	−0.89	——	−0.94	−0.81	——	−0.87	−0.86	——	−0.77

Note: Correlation between Cd and CA content in rice organisms is presented on the lower left and the upper right, respectively. ** indicates extremely significant level ($p < 0.01$), * indicates significant level ($p < 0.05$). —— indicates extremely little or no correlation.

3.1.2. Impacts of Foliar Application of CA on the Speciation of Cd in Various Organs of Rice

Under Cd stress conditions, the insoluble form of Cd was the predominant speciation in various rice organs, and its Cd content far exceeded that of the soluble form. Notably, there was a significant difference in insoluble Cd content versus soluble Cd content in flag leaves, inverted second leaves, and roots, with ratios of approximately 14.23, 17.27, and 9.58, respectively. During the rice flowering stage, the foliar application of CA significantly reduced the soluble Cd content in organs, other than the inverted second leaves and old organizations. Moreover, the reduction effect of 5 mM CA was greater than that of 1 mM CA (Figure 2a). Compared with CK, 5 mM CA led to the following percentage decreases in soluble Cd content in various organs: spikes (53.32%), necks of spikes (39.75%), flag leaves (40.90%), spike nodes (64.44%), inverted second internodes (48.05%), inverted nodes (46.44%), and roots (40.36%), with substantial reductions observed in the neck and panicle nodes. Additionally, when compared with CK, 5 mM CA results in spikes, necks of spikes, flag leaves, inverted second leaves, spike knots, inverted second internodes, inverted nodes, and roots decreased by 36.64%, 26.27%, 35.32%, 38.29%, 55.95%, 43.09%, and 25.21% in insoluble Cd content, with significant reductions primarily observed in spike knots (Figure 2b). Overall, the inhibitory effect of 5 mM CA on soluble Cd content surpassed its effect on insoluble Cd content.

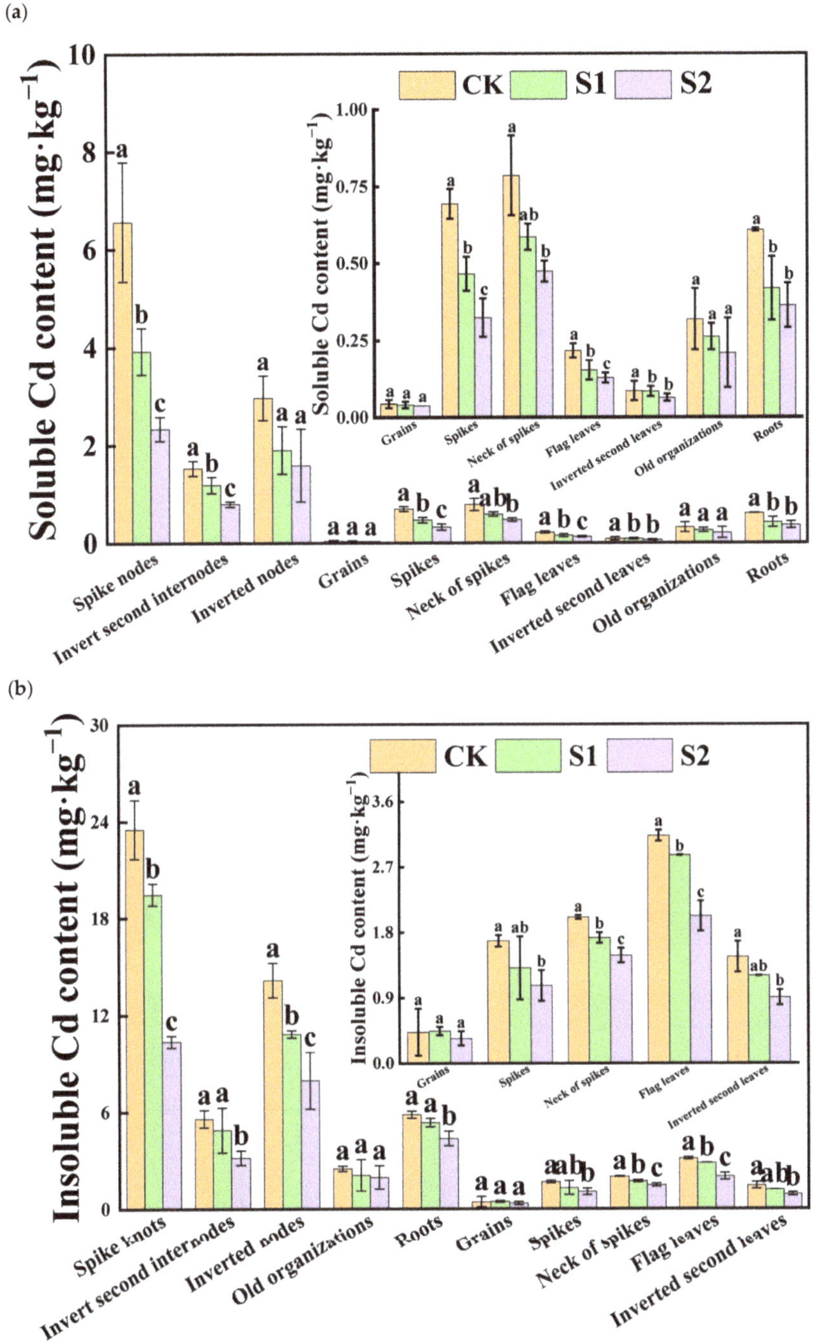

Figure 2. Content of soluble Cd (**a**) and insoluble Cd (**b**) in different tissues of rice. Different letters indicate significant difference ($p < 0.05$) among treatments in the same organ according to Duncan's test.

3.1.3. Effects of CA on Cd Transport between Adjacent Organs in Rice

The foliar application of CA could regulate the distribution ratios of soluble Cd and insoluble Cd in rice organs during the rice flowering stage. The regulatory effect of 5 mM CA surpassed that of 1 mM CA (Figure 3). Specifically, 5 mM CA significantly reduced the distribution ratios of soluble Cd in the spikes, necks of spikes, spike nodes, and roots by 18.52%, 13.33%, 16.02%, and 19.26%, respectively (Figure 3a). Conversely, 5 mM CA increased the distribution ratios of insoluble Cd in the spikes, necks of spikes, spike nodes, and roots by 7.63%, 5.18%, 4.49%, and 2.01%, respectively (Figure 3b).

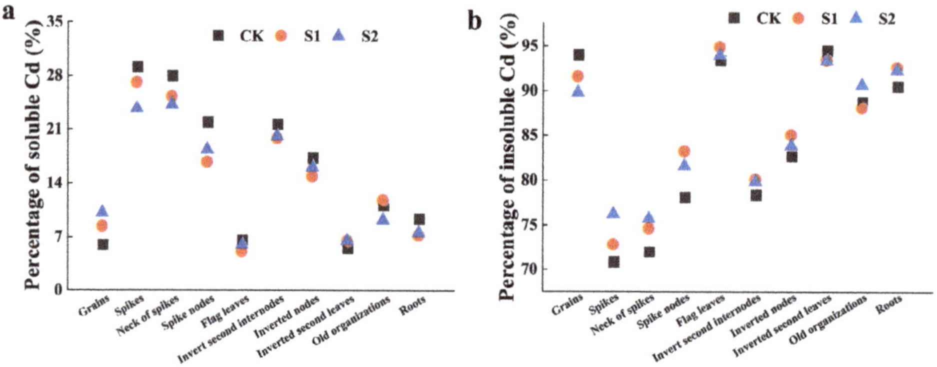

Figure 3. Percentage of soluble Cd (**a**) and insoluble Cd (**b**) in different tissues of rice.

Our experiments found that the transport capacity of soluble Cd to the inverted nodes was relatively strong, with flag leaves exhibiting an intermediate transport capacity to spike nodes. The transport capacity of soluble Cd from the spikes to the grains was weaker, while spike nodes exhibited the weakest transport capacity from soluble Cd to the spikes (Table 2). The spikes, spike nodes, flag leaves, and inverted second leaves played critical roles in the process of Cd transport toward the grains. Both 1 mM and 5 mM CA significantly inhibited the transfer of soluble Cd from the inverted second leaves to the inverted nodes, and 5 mM CA further suppressed the transport of soluble Cd from the flag leaves to the spike nodes. The foliar application of CA did not significantly affect the ability of other organs to transport soluble Cd to adjacent upper organs.

Table 2. Effects of citric acid on transfer factor (TF) of soluble Cd in organs of rice.

	TF Grain/Spike	TF Spike/SK	TF SK/FL	TF ITI/IN	TF IN/ITL	TF ITL/OO	TF OO/Roots
CK	1.3 ± 0.1 a	0.4 ± 0.1 a	141.1 ± 17.7 a	2.5 ± 0.3 a	245.2 ± 44.2 a	8.7 ± 2.3 a	5.0 ± 0.3 a
S1	1.4 ± 0.1 a	0.5 ± 0.1 a	157.7 ± 36.4 a	3.0 ± 0.9 a	155.7 ± 33.3 b	6.5 ± 1.5 a	6.8 ± 2.9 a
S2	1.4 ± 0.3 a	0.6 ± 0.1 a	100.5 ± 13.2 b	2.7 ± 0.7 a	151.9 ± 20.6 b	5.6 ± 1.9 a	6.3 ± 2.0 a

Note: TFA/B = total Cd content in A/soluble Cd content in B. Data represent the means of three repetitions ± standard deviation; lowercase (letters) represent significant differences between different CA treatments ($p < 0.05$). "SK" indicates spike nodes, "FL" indicates flag leaves, "ITI" indicates invert second internodes, "IN" indicates inverted nodes, "ITL" indicates inverted second leaves, "OO" indicates old organizations.

3.2. Essential Mineral Element in Various Organs of Rice

3.2.1. Effects of Foliar Application of CA on Essential Mineral Element Content in Various Organs of Rice

Rice requires abundant essential mineral elements, including K, Ca, and Mg, and trace elements such as Mn, Fe, and Zn. The distribution of mineral elements varies significantly across different rice organs. During the rice flowering stage, the foliar application of CA had the most pronounced impact on the element content in grains, with the effect of 5 mM CA exceeding that of 1 mM CA (Table 3). Among the shoots, spike nodes consistently exhibited the highest element content, while grains had the lowest. Notably, the Fe in roots far exceeded that in other organs. Compared with CK, 5 mM CA significantly increased K, Mg, and Mn content in grains by 46.40%, 82.40%, and 18.93%, respectively, while reducing Ca and Zn content by 23.19% and 14.13%, respectively. Additionally, 5 mM CA significantly enhanced K in the spikes by 17.43% and Fe by 1.87 times. It also increased Mn content in spike knots by 23.99% but decreased Zn content by 17.03%. Furthermore, 5 mM CA led to a substantial increase in Mn content in the roots by 50.37%.

Table 3. Effects of CA on essential mineral elements in tissues of rice.

Organ	Treatment	K ($g \cdot kg^{-1}$)	Ca ($g \cdot kg^{-1}$)	Mg ($g \cdot kg^{-1}$)	Mn ($g \cdot kg^{-1}$)	Fe ($g \cdot kg^{-1}$)	Zn ($g \cdot kg^{-1}$)
Grains	CK	0.87 ± 0.03 b	0.57 ± 0.05 a	0.32 ± 0.01 c	0.02 ± 0 b	0.02 ± 0 a	0.02 ± 0 a
	S1	1.04 ± 0.21 b	0.45 ± 0.02 b	0.48 ± 0.05 b	0.02 ± 0 ab	0.02 ± 0 a	0.02 ± 0 b
	S2	1.27 ± 0.05 a	0.44 ± 0.02 b	0.57 ± 0.03 a	0.02 ± 0 a	0.02 ± 0 a	0.02 ± 0 b
Spikes	CK	21.08 ± 1.29 b	1.74 ± 0.05 a	1 ± 0.07 a	0.38 ± 0.03 a	0.13 ± 0.02 b	0.06 ± 0.01 a
	S1	26.44 ± 0.8 a	1.85 ± 0.37 a	1.06 ± 0.11 a	0.35 ± 0.05 a	0.27 ± 0.02 a	0.07 ± 0.01 a
	S2	24.761.05 a	2.12 ± 0.22 a	1.05 ± 0.08 a	0.38 ± 0.03 a	0.36 ± 0.07 a	0.08 ± 0.01 a
Spike nodes	CK	133.92 ± 15.42 a	7.92 ± 0.43 a	5.85 ± 0.89 a	1.42 ± 0.06 b	1.24 ± 0.09 a	1.37 ± 0.06 a
	S1	139.2 ± 26.8 a	7.56 ± 0.24 a	5.42 ± 0.22 a	1.62 ± 0.11 ab	1.44 ± 0.12 a	1.19 ± 0.1 b
	S2	128.76 ± 13.93 a	7.03 ± 0.64 a	4.99 ± 0.49 a	1.76 ± 0.20 a	1.48 ± 0.25 a	1.14 ± 0.01 b
Flag leaves	CK	19.32 ± 1.82 a	5.64 ± 0.26 a	2.38 ± 0.41 a	1.26 ± 0.18 a	0.29 ± 0.03 a	0.05 ± 0.01 a
	S1	21.49 ± 2.68 a	6.17 ± 0.88 a	2.46 ± 0.37 a	1.27 ± 0.08 a	0.33 ± 0.05 a	0.05 ± 0.01 a
	S2	20.28 ± 2.27 a	6.48 ± 0.71 a	2.41 ± 0.45 a	1.14 ± 0.1 a	0.4 ± 0.06 a	0.06 ± 0.01 a
Others	CK	33.47 ± 0.63 a	6.91 ± 0.8 a	2.51 ± 0.2 a	1.13 ± 0.08 a	0.67 ± 0.05 a	0.1 ± 0.01 a
	S1	34.02 ± 0.16 a	7.19 ± 0.15 a	2.44 ± 0.03 a	1.15 ± 0.04 a	0.84 ± 0.07 a	0.09 ± 0.01 a
	S2	31.76 ± 4.82 a	7.15 ± 1.13 a	2.44 ± 0.16 a	0.93 ± 0.1 b	0.87 ± 0.16 a	0.08 ± 0.01 a
Roots	CK	6.55 ± 1.11 a	4.78 ± 0.73 a	1.15 ± 0.19 a	0.66 ± 0.12 b	75.78 ± 6.55 a	0.1 ± 0.02 a
	S1	7.61 ± 1.7 a	4.67 ± 0.07 a	1.19 ± 0.02 a	0.51 ± 0.07 b	67.09 ± 12.85 a	0.09 ± 0.01 a
	S2	5.86 ± 1.4 a	5.65 ± 0.81 a	1.24 ± 0.21 a	0.99 ± 0.06 a	81.42 ± 13.11 a	0.11 ± 0.02 a

Note: Different letters indicate significant difference among treatments in the same organ and same element according to ANOVA analysis of variance ($p < 0.05$). Data represent the means of three repetitions ± standard deviation.

3.2.2. Effects of Foliar Application of CA on the Speciation of Mn in Rice Spike Knots and Roots

In both rice spike knots and roots, the content of insoluble Mn was higher than that of soluble Mn (Figure 4). Under Cd stress conditions, the insoluble Mn content in spike nodes was approximately 2.41 times higher compared with soluble Mn, while, in roots, the insoluble Mn content was 7.82 times higher than soluble Mn. The foliar application of CA could modulate the speciation of Mn within these organs during the rice flowering stage. An amount of 1 mM CA had no significant impact on the soluble Mn content in rice roots and spike nodes, as shown in Figure 4. However, compared with CK, 5 mM CA significantly increased the soluble Mn content in roots by 27.18% and in spike nodes by 42.06%. Neither 1 mM nor 5 mM CA significantly affected the insoluble Mn content in rice spike nodes. However, 5 mM CA substantially enhanced the insoluble Mn content in roots, with an impressive increase of 84.48%.

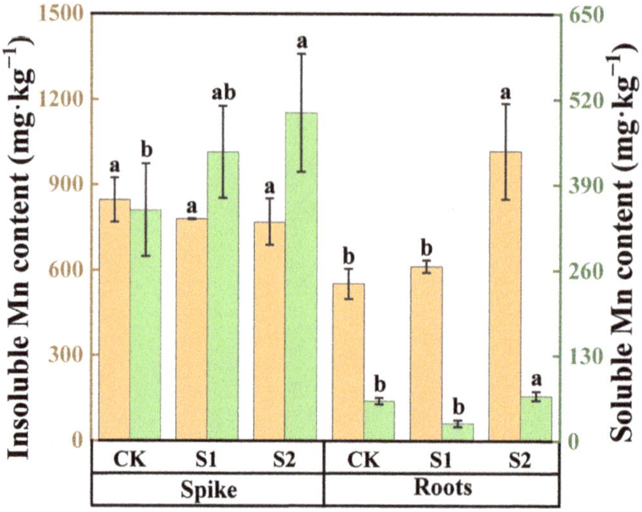

Figure 4. Content of soluble and insoluble Mn in spikes and roots of rice treated with different concentrations of CA. Data represent the means of three repetitions ± standard deviation and lowercase letters represent significant differences between different CA treatments ($p < 0.05$).

3.3. Effects of Foliar Application of CA on Ion Balance in Key Organs of Rice

3.3.1. Effects of Foliar Application of CA on Mn:Cd Ratios in Different Organs

The ratios of Mn content to soluble Cd content in flag leaves was significantly higher than in spikes, spike nodes, and roots, with the smallest ratios observed in spike nodes (Figure 5). During the rice flowering stage, the foliar application of 1 mM CA significantly increased the Mn:Cd ratios in spike nodes by 87.68%. Moreover, 5 mM CA led to substantial ratio increases of 104.62% and 235.08% in both spikes and spike nodes (Figure 5a). Additionally, 1 mM CA significantly enhanced the ratios in flag leaves by 45.12%, while 5 mM CA significantly increased the Mn:Cd ratios in different organs in both flag leaves and roots; the increases were 74.20% and 160.12%, respectively (Figure 5b).

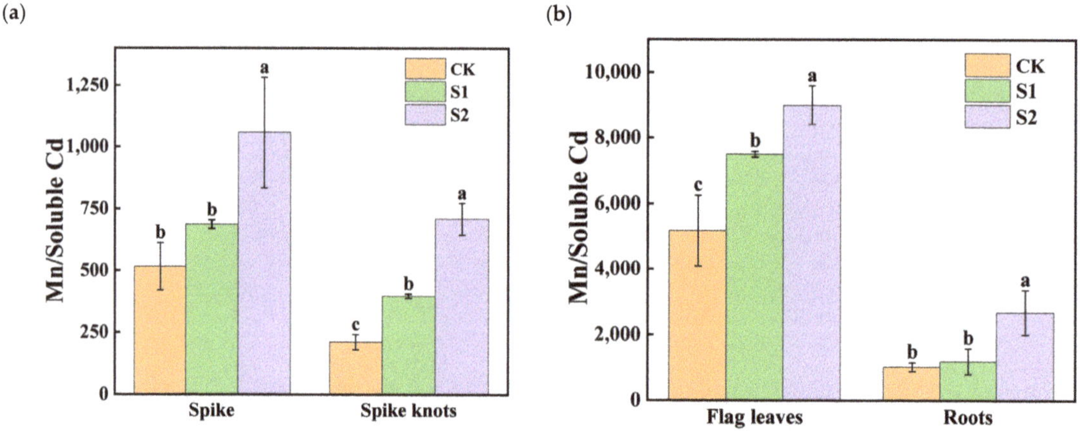

Figure 5. Effect of different concentrations of CA on Mn/soluble Cd in spikes and spike nodes (**a**) and flag leaves and roots (**b**) of rice. Data represent the means of three repetitions ± standard deviation and lowercase letters represent significant differences between different CA treatments ($p < 0.05$).

3.3.2. Effects of Foliar Application of CA on the Ratio (Ca:Mn) to Soluble Cd Ratio in Different Organs

The ratio of (Ca:Mn) to soluble Cd in various rice organs from flag leaves, old organizations, roots to spike nodes showed a gradually decreasing trend. During the rice flowering stage, the foliar application of 1 mM CA significantly increased this ratio in all organs by 25.63% to 77.04% compared with CK. Additionally, 5 mM CA led to substantial increases in the ratio of (Ca:Mn) to soluble Cd in both spikes and spike nodes, increasing by 161.45% and 95.12% compared with CK (Figure 6). Furthermore, it significantly enhanced this ratio in flag leaves and old organizations when compared with CK, increasing by 146.59% and 153.45%, respectively.

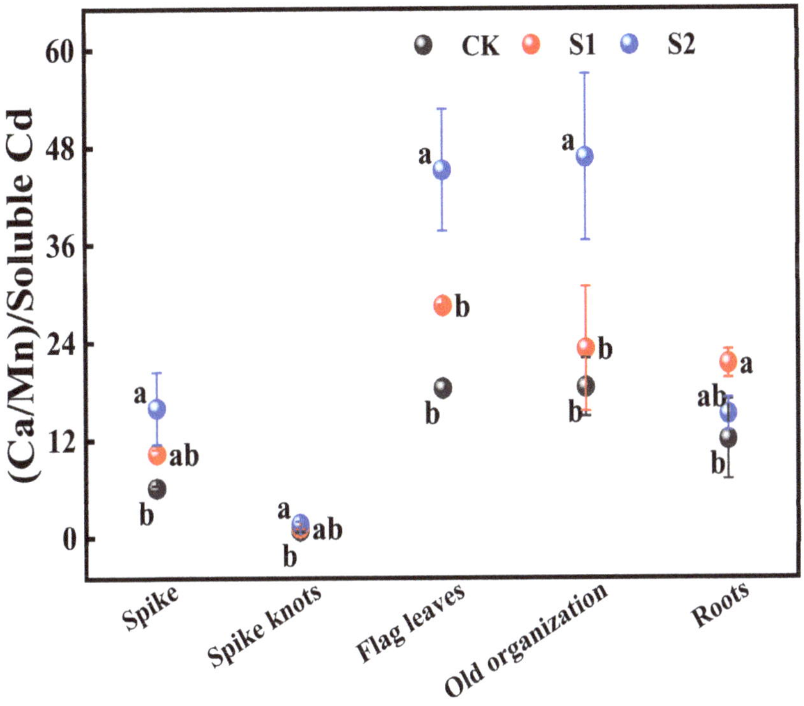

Figure 6. Impact of different concentrations of CA on (Ca:Mn)/soluble Cd in different key organs of rice. Data represent the means of three repetitions ± standard deviation and lowercase letters represent significant differences between different CA treatments ($p < 0.05$).

3.4. Correlation Analysis

CA could affect the accumulation of Cd in different organs of rice and had a good effect on Cd reduction, and the content of Cd in different organs of rice had a significant effect on the content of mineral elements (Figure 7). Correlation analyses showed that soluble Cd was negatively correlated with the contents of Ca and Mn in different organs of rice, and the differences reached highly significant levels. Soluble Cd was significantly and negatively correlated with Mg and Mn:Cd in different organs of rice, and insoluble Cd was significantly and positively correlated with Fe content and (Ca:Mn). CA content in grains, spikes, spike nodes, flag leaves, inverted second internodes, and roots were positively correlated with Zn, Mn, Mg, and K content, and significantly negatively correlated with Fe and (Ca:Mn).

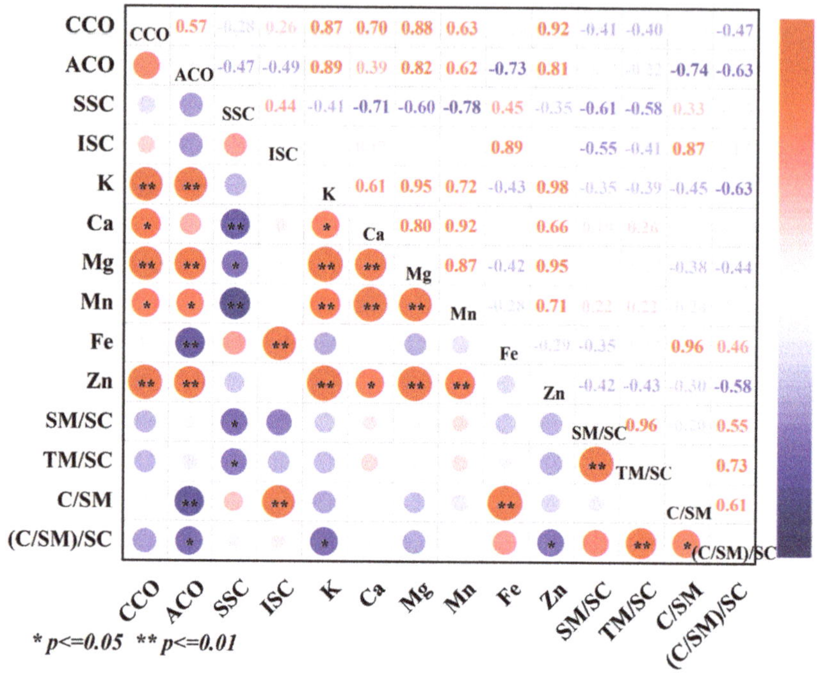

Figure 7. Correlation analysis among the treatments. Data represent the mean of three replicates ± standard deviation. "CCO" indicates Cd content in organs, "ACO" indicates CA content in organs, "SSC" indicates soluble Cd content, "ISC" indicates insoluble Cd content, "SM/SC" indicates soluble Mn/soluble Cd, "TM/SC" indicates total Mn/soluble Cd, "C/SM" indicates Ca/soluble Mn, "(C/SM)/SC" indicates (Ca/soluble Mn)/soluble Cd.

4. Discussion

4.1. Foliar Application of CA Could Reduce Rice Absorption and Transport of Cd

Pairs of hydroxyl and carboxyl/carboxyl and carboxyl groups attached to neighboring carbon atoms in CA can form stable five-membered or six-membered cyclic structures by chelating heavy metals within the plant [38]. Under stress conditions, plants may sequester heavy metals in vacuoles, where most heavy metals could be chelated with compounds like CA, thus reducing toxicity risks [39,40]. Research has shown that the addition of organic acid supplements could decrease Cd content in vacuoles, and, through enhancing the isolation of Cd by root vacuoles, thus contribute to maintaining plant growth and reducing Cd content [17]. The foliar application of inhibitors could reduce the transport of Cd from rice roots and stem bases to other parts by altering grain quality and lowering rice Cd content [41]. Additionally, when faced with heavy metal stress, CA may protect plant cell-structure integrity and stability by restoring ion balance within the plant. It enhanced cell membrane selectivity for heavy metal ions and formed stable ligand complexes that prevented metal absorption by roots, thus avoiding metal accumulation at sensitive sites in the roots [8,42]. Due to its ability to chelate heavy metals and promote plant growth, bean genotypes with higher secretion levels of CA exhibited lower aluminum (Al) content per unit in their grains [43]. Exposed to heavy metal stress, the foliar application of thiol compounds could inhibit the migration of Cd, Pb, and As within rice flag leaves, thereby reducing the accumulation of relevant heavy metals in grains [44]. This suggests that foliar spraying methods may enhance the metal-blocking capacity of specific rice organs, leading to a decrease in Cd content in rice grains.

Our research findings indicated that applying CA during the rice flowering stage increased CA content in rice organs such as spikes, spike nodes, flag leaves, inverted nodes, inverted second leaves, and roots. The effect was even better with 5 mM CA compared with 1 mM CA. CA could also reduce Cd content in tissues other than roots and old organizations. Notably, the content of CA in these tissues was significantly negatively correlated with Cd content. Specifically, 5 mM CA significantly reduced Cd content in grains and spikes by 52% and 37%, respectively. This suggests that the foliar application of CA may achieve Cd toxicity reduction by increasing the levels of CA in different rice organs, altering the chemical form of Cd, and chelating and sequestering Cd within these organs, thereby limiting Cd migration to rice grains

Rice organs could convert most of the Cd that entered the plant into insoluble forms, thereby preventing excessive Cd transport to the grains. This metal-blocking effect reached its limit when Cd stress concentrations were high [20]. Research has shown that external applications of CA could convert more Cd in radish stems and leaves into insoluble forms [45]. Our study found that the foliar application of 5 mM CA significantly reduced the soluble Cd content in nutrient-rich organs, other than inverted two leaves and old organizations, during the rice flowering stage. Additionally, it markedly decreased the proportion of soluble Cd in spikes, necks of spikes, spike nodes, and roots, while increasing the proportion of insoluble Cd in these organs and then lowering the transfer factor for Cd transport from flag leaves to spikes and from inverted second leaves to inverted nodes. This indicates that the foliar application of CA could enhance the metal-blocking capacity of rice nutrient-rich organs during the rice flowering stage and reduce the transport ability of Cd within the rice plant.

4.2. Foliar Application of CA Could Alleviate Cd Content by Increasing the Uptake of Mineral Elements and Reducing the Content of Soluble Cd in Different Organs of Rice

Different elemental ions exhibit synergistic or antagonistic effects. During the growth and development stage of rice, the plant regulates the absorption of essential elements to enhance its ability to withstand external disturbances and maintain normal growth [46]. Under heavy metal stress, the capacity of rice to absorb most essential elements from the roots would be weakened. To sustain normal growth, rice adjusts the levels of elements such as Ca, K, Mg, and Mn to maintain proper antioxidant levels [47]. For instance, when exposed to Cd stress, *Arabidopsis* modulates Ca levels to maintain normal antioxidant systems, with Ca acting as a secondary messenger [48]. Research has identified the FBP (fructose bisphosphatase) gene's role in mediated antagonistic interactions between certain metal ions within plants. When Fe levels are low and Zn levels are high, FBP expression decreases while NAS (nicotianamine synthase) expression increases, which chelates Fe and Zn and then increases its plant availability. Simultaneously, it reduces Zn transport to the aerial parts, preventing Zn toxicity [49]. In this study, we found that Cd content in various organs of rice was significantly positively correlated with K, Ca, Mg, Mn, and Zn, and Cd content in various organs was significantly reduced under CA treatment, indicating that the application of CA reduced the absorption capacity of Cd in rice and improved the absorption of other mineral elements to maintain the normal growth.

Transport proteins in different nodes of rice mediate the distribution of mineral elements within grains, thereby regulating nutrient uptake, translocation, and distribution [50]. *OsNramp3* (natural resistance-associated macrophage protein) is closely associated with the transport of Mn in rice, and the expression of this transporter-related gene in the xylem and cortex promotes the preferential transfer of Mn to parts such as flag leaves. *OsLCT1* (low-affinity cation transporter) may participate in the relocalization of other elements within leaves, influencing the transport of Cd to rice grains [50,51]. These mechanisms may contribute to the variations in nutrient elements observed in different rice organs under Cd stress.

Acting as a metabolite signal, CA has been shown to play a crucial role in regulating various mineral elements [52–54]. Increased CA concentration in soil enhances the ability of plants to cope with Fe deficiency [55]. Not only does CA provide additional anions to cells but it also competes with intracellular phosphate ions for the adsorption of elements like Mn, thereby enhancing their plant availability. At the same time, the binding of phosphate ions with Cd is strengthened, further inhibiting the activity of Cd [56,57]. Our study reveals that, during the rice flowering stage, the foliar application of 5 mM CA significantly increases soluble and insoluble Mn content in grains, spike nodes, and roots; increased Mn content significantly reduces soluble Cd content in rice seeds, spikes, spike nodes, flag leaves, inverted second internodes, and roots, thus reducing Cd transport to various organs and alleviating Cd toxicity.

4.3. Foliar Application of CA Alleviates Rice Organ Cd Absorption by Modulating Ca:Mn Ratios

Clarifying the interactions between elements with plant tissues may be an effective solution to mitigate the toxicity of Cd in rice. Currently, the presence of Ca and Mn has been found to reduce the transfer capacity of Cd from roots to shoots and from shoots to grains in cropland soil, thereby mitigating rice grain Cd contamination. Additionally, Cd content is typically positively correlated with Mn content in rice organs [58,59]. Knocking out *OsNramp5* simultaneously reduces Mn and Cd content in different rice organs [60]. Under Cd stress conditions, nutrient-rich organs could alleviate Cd toxicity by adjusting the Ca: Mn ratios. High ratios of Ca: Mn are associated with low Cd accumulation in these organs [20]. Our study reveals that the foliar application of 5 mM CA significantly increases the Mn and soluble Cd ratios in spikes, spike nodes, flag leaves, and roots during the rice flowering stage. This adjustment improves the physiological process of simultaneously absorbing Mn and Cd in rice. Furthermore, it significantly enhances the (Ca:Mn) ratio in spikes, spike nodes, flag leaves, old organizations, and roots in terms of soluble Cd content. At the same time, the increase in (Ca:Mn) would increase the content of insoluble Cd in various organs of rice, reducing the transfer of Cd to rice seeds, and effectively inhibiting rice Cd accumulation.

5. Conclusions

During the flowering stage of rice, spraying CA could increase the content of CA in rice organs and reduce the Cd content in tissues other than the root system and old organizations. Compared with 1 mM CA, 5 mM CA made Cd content in grains and spikes significantly decrease by 52% and 37%, respectively. Additionally, 5 mM CA significantly reduced the soluble Cd content in nutrient organs other than the inverted second leaves and old organizations, and in spikes, necks of spikes, spike nodes, and roots. This reduction in soluble Cd allocation was accompanied by an increase in the allocation of insoluble Cd in these organs. Furthermore, the transfer factor for Cd transport from flag leaves to spike nodes and from the inverted second leaves to the inverted nodes decreased, indicating that the inhibitory effect of rice vegetative organs on Cd migration and transport in rice increased. An amount of 5 mM CA also significantly increased the content of Mn in grains, spike nodes, and roots by 21%, 24%, and 33%, respectively. It led to a substantial increase in soluble Mn content in spike nodes and roots by 42% and 20%, respectively, while significantly reducing the content of soluble Cd in rice grain, spikes, spike nodes, flag leaves, inverted second internodes, and roots. Moreover, the ratios of Mn to soluble Cd in spikes, spike nodes, flag leaves, and roots improved the antagonistic effect of Mn and Cd by rice during physiological processes. And the ratio of (Ca:Mn) to soluble Cd in the spikes, spike knots, flag leaves, old organizations, and roots significantly increased. In summary, the foliar application of CA on farmland could decrease the content of soluble Cd, increase the Mn content and the ratios of Ca to Mn of different vegetative organs of rice, and increase the ability of vegetative organs to convert Cd into insoluble Cd, thus reducing the transfer of Cd to the rice grain.

Author Contributions: Conceptualization, K.C.; funding acquisition, W.X.; investigation, B.Y.; methodology, K.C., B.Y., X.G., X.Z., J.Y. and B.Z.; project administration, W.X.; supervision, Y.S.;

writing—original draft, K.C.; writing—review and editing, W.X., C.Z. and Y.D. All authors have read and agreed to the published version of the manuscript.

Funding: This work was supported by the National Key Research and Development Program of China (No. 2022YFD1700103), National Natural Science Foundation of China (42207039), Funds for Science and Technology Innovation Project from the Chinese Academy of Agricultural Sciences, and the Central Public Research Institute Basic Fund for Research and Development.

Institutional Review Board Statement: Not applicable.

Informed Consent Statement: Not applicable.

Data Availability Statement: Data set available on request from the authors.

Acknowledgments: Thanks to all the people who have helped to write and revise the article.

Conflicts of Interest: The authors declare no conflicts of interest.

References

1. Zhang, X.; Zhong, T.; Liu, L.; Ouyang, X. Impact of Soil Heavy Metal Pollution on Food Safety in China. *PLoS ONE* **2015**, *10*, e0135182. [CrossRef]
2. Hamid, Y.; Tang, L.; Sohail, M.I.; Cao, X.; Hussain, B.; Aziz, M.Z.; Usman, M.; He, Z.; Yang, X. An Explanation of Soil Amendments to Reduce Cadmium Phytoavailability and Transfer to Food Chain. *Sci. Total Environ.* **2019**, *660*, 80–96. [CrossRef] [PubMed]
3. Ali, H.; Khan, E. What Are Heavy Metals? Long-Standing Controversy over the Scientific Use of the Term 'Heavy Metals'—Proposal of a Comprehensive Definition. *Toxicol. Environ. Chem.* **2018**, *100*, 6–19. [CrossRef]
4. Li, J.; Xu, Y. Immobilization Remediation of Cd-Polluted Soil with Different Water Condition. *J. Environ. Manag.* **2017**, *193*, 607–612. [CrossRef] [PubMed]
5. Peera Sheikh Kulsum, P.G.; Khanam, R.; Das, S.; Nayak, A.K.; Tack, F.M.G.; Meers, E.; Vithanage, M.; Shahid, M.; Kumar, A.; Chakraborty, S.; et al. A State-of-the-Art Review on Cadmium Uptake, Toxicity, and Tolerance in Rice: From Physiological Response to Remediation Process. *Environ. Res.* **2023**, *220*, 115098. [CrossRef] [PubMed]
6. Riaz, M.; Kamran, M.; Rizwan, M.; Ali, S.; Wang, X. Foliar Application of Silica Sol Alleviates Boron Toxicity in Rice (*Oryza sativa*) Seedlings. *J. Hazard. Mater.* **2022**, *423*, 127175. [CrossRef] [PubMed]
7. Wang, C.; Rong, H.; Zhang, X.; Shi, W.; Hong, X.; Liu, W.; Cao, T.; Yu, X.; Yu, Q. Effects and Mechanisms of Foliar Application of Silicon and Selenium Composite Sols on Diminishing Cadmium and Lead Translocation and Affiliated Physiological and Biochemical Responses in Hybrid Rice (*Oryza sativa* L.) Exposed to Cadmium and Lead. *Chemosphere* **2020**, *251*, 126347. [CrossRef] [PubMed]
8. Ehsan, S.; Ali, S.; Noureen, S.; Mahmood, K.; Farid, M.; Ishaque, W.; Shakoor, M.B.; Rizwan, M. Citric Acid Assisted Phytoremediation of Cadmium by *Brassica napus* L. *Ecotoxicol. Environ. Saf.* **2014**, *106*, 164–172. [CrossRef] [PubMed]
9. Freitas, E.V.; Nascimento, C.W.; Souza, A.; Silva, F.B. Citric Acid-Assisted Phytoextraction of Lead: A Field Experiment. *Chemosphere* **2013**, *92*, 213–217. [CrossRef]
10. Nazari, S.; Rahimi, G.; Khademi Jolgeh Nezhad, A. Effectiveness of Native and Citric Acid-Enriched Biochar of Chickpea Straw in Cd and Pb Sorption in an Acidic Soil. *J. Environ. Chem. Eng.* **2019**, *7*, 103064. [CrossRef]
11. Kibria, M.G.; Barton, L.; Rengel, Z. Applying Foliar Magnesium Enhances Wheat Growth in Acidic Soil by Stimulating Exudation of Malate and Citrate. *Plant Soil* **2021**, *464*, 621–634. [CrossRef]
12. Patra, D.K.; Pradhan, C.; Patra, H.K. Chelate Based Phytoremediation Study for Attenuation of Chromium Toxicity Stress Using Lemongrass: *Cymbopogon flexuosus* (Nees Ex Steud.) W. Watson. *Int. J. Phytoremediation* **2018**, *20*, 1324–1329. [CrossRef] [PubMed]
13. Ishikawa, S.; Wagatsuma, T.; Sasaki, R.; Ofei-Manu, P. Comparison of the Amount of Citric and Malic Acids in Al Media of Seven Plant Species and Two Cultivars Each in Five Plant Species. *Soil Sci. Plant Nutr.* **2000**, *46*, 751–758. [CrossRef]
14. Zeng, F.; Chen, S.; Miao, Y.; Wu, F.; Zhang, G. Changes of Organic Acid Exudation and Rhizosphere pH in Rice Plants under Chromium Stress. *Environ. Pollut.* **2008**, *155*, 284–289. [CrossRef]
15. Ben Massoud, M.; Sakouhi, L.; Chaoui, A. Effect of Plant Growth Regulators, Calcium and Citric Acid on Copper Toxicity in Pea Seedlings. *J. Plant Nutr.* **2019**, *42*, 1230–1242. [CrossRef]
16. Liu, G.; Meng, J.; Zeng, L.; Liu, X.; Dai, Z.; Tang, C.; Xu, J. Novel Agricultural Waste-Based Materials Decrease the Uptake and Accumulation of Cadmium by Rice (*Oryza sativa* L.) in Contaminated Paddy Soils. *Environ. Pollut.* **2021**, *289*, 117838. [CrossRef]
17. Sebastian, A.; Prasad, M.N.V. Exogenous Citrate and Malate Alleviate Cadmium Stress in *Oryza sativa* L.: Probing Role of Cadmium Localization and Iron Nutrition. *Ecotoxicol. Environ. Saf.* **2018**, *166*, 215–222. [CrossRef] [PubMed]
18. Li, F.; Chen, X.; Feng, J.; Liang, Z.; Xu, X.; Ding, T. Ryegrass Extraction of Heavy Metals from Municipal Sewage Sludge Compost-Amended Soils Assisted with Citric Acid. *Environ. Sci. Pollut. Res.* **2023**, *30*, 33598–33608. [CrossRef]
19. Zhang, X.; Zhong, B.; Shafi, M.; Guo, J.; Liu, C.; Guo, H.; Peng, D.; Wang, Y.; Liu, D. Effect of EDTA and Citric Acid on Absorption of Heavy Metals and Growth of Moso Bamboo. *Environ. Sci. Pollut. Res.* **2018**, *25*, 18846–18852. [CrossRef]
20. Xue, W.-J.; Zhang, C.-B.; Wang, P.-P.; Wang, C.-R.; Huang, Y.-C.; Zhang, X.; Liu, Z.-Q. Rice Vegetative Organs Alleviate Cadmium Toxicity by Altering the Chemical Forms of Cadmium and Increasing the Ratio of Calcium to Manganese. *Ecotoxicol. Environ. Saf.* **2019**, *184*, 109640. [CrossRef]

21. Li, S.; Li, G.; Huang, X.; Chen, Y.; Lv, C.; Bai, L.; Zhang, K.; He, H.; Dai, J. Cultivar-Specific Response of Rhizosphere Bacterial Community to Uptake of Cadmium and Mineral Elements in Rice (*Oryza sativa* L.). *Ecotoxicol. Environ. Saf.* **2023**, *249*, 114403. [CrossRef]
22. Ishimaru, Y.; Takahashi, R.; Bashir, K.; Shimo, H.; Senoura, T.; Sugimoto, K.; Ono, K.; Yano, M.; Ishikawa, S.; Arao, T.; et al. Characterizing the Role of Rice NRAMP5 in Manganese, Iron and Cadmium Transport. *Sci. Rep.* **2012**, *2*, 286. [CrossRef]
23. Carrió-Seguí, A.; Garcia-Molina, A.; Sanz, A.; Peñarrubia, L. Defective Copper Transport in the Copt5 Mutant Affects Cadmium Tolerance. *Plant Cell Physiol.* **2015**, *56*, 442–454. [CrossRef]
24. Cho, S.-C.; Chao, Y.-Y.; Kao, C.H. Calcium Deficiency Increases Cd Toxicity and Ca Is Required for Heat-Shock Induced Cd Tolerance in Rice Seedlings. *J. Plant Physiol.* **2012**, *169*, 892–898. [CrossRef]
25. Chou, T.-S.; Chao, Y.-Y.; Huang, W.-D.; Hong, C.-Y.; Kao, C.H. Effect of Magnesium Deficiency on Antioxidant Status and Cadmium Toxicity in Rice Seedlings. *J. Plant Physiol.* **2011**, *168*, 1021–1030. [CrossRef] [PubMed]
26. Yang, J.; Sun, H.; Qin, J.; Wang, X.; Chen, W. Impacts of Cd on Temporal Dynamics of Nutrient Distribution Pattern of Bletilla Striata, a Traditional Chinese Medicine Plant. *Agriculture* **2021**, *11*, 594. [CrossRef]
27. Xue, W.; Wang, P.; Tang, L.; Zhang, C.; Wang, C.; Huang, Y.; Zhang, X.; Li, Y.; Zhao, B.; Liu, Z. Citric Acid Inhibits Cd Uptake by Improving the Preferential Transport of Mn and Triggering the Defense Response of Amino Acids in Grains. *Ecotoxicol. Environ. Saf.* **2021**, *211*, 111921. [CrossRef] [PubMed]
28. You, S.; Deng, Z.; Chen, M.; Zheng, Y.; Liu, J.; Jiang, P. Mn Pretreatment Improves the Physiological Resistance and Root Exudation of Celosia Argentea Linn. to Cadmium Stress. *Int. J. Environ. Res. Public Health* **2023**, *20*, 1065. [CrossRef]
29. da Silva, J.T.; Paniz, F.P.; Sanchez, F.E.S.; Pedron, T.; Torres, D.P.; da Rocha Concenço, F.I.G.; Barbat Parfitt, J.M.; Batista, B.L. Selected Soil Water Tensions at Phenological Phases and Mineral Content of Trace Elements in Rice Grains—Mitigating Arsenic by Water Management. *Agric. Water Manag.* **2020**, *228*, 105884. [CrossRef]
30. Zhou, H.; Zhu, W.; Yang, W.-T.; Gu, J.-F.; Gao, Z.-X.; Chen, L.-W.; Du, W.-Q.; Zhang, P.; Peng, P.-Q.; Liao, B.-H. Cadmium Uptake, Accumulation, and Remobilization in Iron Plaque and Rice Tissues at Different Growth Stages. *Ecotoxicol. Environ. Saf.* **2018**, *152*, 91–97. [CrossRef]
31. Wang, S.; Wang, F.; Gao, S. Foliar Application with Nano-Silicon Alleviates Cd Toxicity in Rice Seedlings. *Environ. Sci. Pollut. Res. Int.* **2015**, *22*, 2837–2845. [CrossRef] [PubMed]
32. Guo, T.; Wei, J.; Li, X.; Yu, J. Environmental Context of Phenotypic Plasticity in Flowering Time in Sorghum and Rice. *J. Exp. Bot.* **2024**, *75*, 1004–1015. [CrossRef] [PubMed]
33. Ishimaru, T.; Hlaing, K.T.; Oo, Y.M.; Lwin, T.M.; Sasaki, K.; Lumanglas, P.D.; Simon, E.-V.M.; Myint, T.T.; Hairmansis, A.; Susanto, U.; et al. An Early-Morning Flowering Trait in Rice Can Enhance Grain Yield under Heat Stress Field Conditions at Flowering Stage. *Field Crops Res.* **2022**, *277*, 108400. [CrossRef]
34. Zhao, M.; Qian, E.; Zhang, F.; Liu, R.; Liu, X.; Zhao, Y.; Liang, X. Spatiotemporal Dynamics of Labile Cd in Soil during Rice Growth. *Sci. Total Environ.* **2020**, *738*, 139832. [CrossRef] [PubMed]
35. Fu, X.; Dou, C.; Chen, Y.; Chen, X.; Shi, J.; Yu, M.; Xu, J. Subcellular Distribution and Chemical Forms of Cadmium in *Phytolacca americana* L. *J. Hazard. Mater.* **2011**, *186*, 103–107. [CrossRef] [PubMed]
36. Yang, L.; Zeng, J.; Wang, P.; Zhu, J. Sodium Hydrosulfide Alleviates Cadmium Toxicity by Changing Cadmium Chemical Forms and Increasing the Activities of Antioxidant Enzymes in Salix. *Environ. Exp. Bot.* **2018**, *156*, 161–169. [CrossRef]
37. Gao, M.; Zhou, J.; Liu, H.; Zhang, W.; Hu, Y.; Liang, J.; Zhou, J. Foliar Spraying with Silicon and Selenium Reduces Cadmium Uptake and Mitigates Cadmium Toxicity in Rice. *Sci. Total Environ.* **2018**, *631–632*, 1100–1108. [CrossRef]
38. Xue, W.; Zhang, X.; Zhang, C.; Wang, C.; Huang, Y.; Liu, Z. Mitigating the Toxicity of Reactive Oxygen Species Induced by Cadmium via Restoring Citrate Valve and Improving the Stability of Enzyme Structure in Rice. *Chemosphere* **2023**, *327*, 138511. [CrossRef]
39. DalCorso, G.; Farinati, S.; Furini, A. Regulatory Networks of Cadmium Stress in Plants. *Plant Signal. Behav.* **2010**, *5*, 663–667. [CrossRef]
40. Martinoia, E. Vacuolar Transporters—Companions on a Longtime Journey. *Plant Physiol.* **2018**, *176*, 1384–1407. [CrossRef]
41. Li, B.; Wang, S.; You, X.; Wen, Z.; Huang, G.; Huang, C.; Li, Q.; Chen, K.; Zhao, Y.; Gu, M.; et al. Effect of Foliar Spraying of Gibberellins and Brassinolide on Cadmium Accumulation in Rice. *Toxics* **2023**, *11*, 364. [CrossRef]
42. Ingelmo, F.; Molina, M.J.; Soriano, M.D.; Gallardo, A.; Lapeña, L. Influence of Organic Matter Transformations on the Bioavailability of Heavy Metals in a Sludge Based Compost. *J. Environ. Manag.* **2012**, *95*, S104–S109. [CrossRef] [PubMed]
43. Mugai, E.N.; Agong, S.G.; Matsumoto, H. Aluminium Tolerance Mechanisms in *Phaseolus vulgaris* L.: Citrate Synthase Activity and TTC Reduction Are Well Correlated with Citrate Secretion. *Soil. Sci. Plant Nutr.* **2000**, *46*, 939–950. [CrossRef]
44. Yang, X.; Wang, C.; Huang, Y.; Liu, B.; Liu, Z.; Huang, Y.; Cheng, L.; Huang, Y.; Zhang, C. Foliar Application of the Sulfhydryl Compound 2,3-Dimercaptosuccinic Acid Inhibits Cadmium, Lead, and Arsenic Accumulation in Rice Grains by Promoting Heavy Metal Immobilization in Flag Leaves. *Environ. Pollut.* **2021**, *285*, 117355. [CrossRef]
45. Ibrahim, E.A. Effect of Citric Acid on Phytoextraction Potential of Cucurbita Pepo, Lagenaria Siceraria, and Raphanus Sativus Plants Exposed to Multi-Metal Stress. *Sci. Rep.* **2023**, *13*, 13070. [CrossRef] [PubMed]
46. Huang, S.; Wang, P.; Yamaji, N.; Ma, J.F. Plant Nutrition for Human Nutrition: Hints from Rice Research and Future Perspectives. *Mol. Plant* **2020**, *13*, 825–835. [CrossRef]

47. Zhu, Y.; Wu, Q.; Lv, H.; Chen, W.; Wang, L.; Shi, S.; Yang, J.; Zhao, P.; Li, Y.; Christopher, R.; et al. Toxicity of Different Forms of Antimony to Rice Plants: Effects on Reactive Oxidative Species Production, Antioxidative Systems, and Uptake of Essential Elements. *Environ. Pollut.* **2020**, *263*, 114544. [CrossRef] [PubMed]
48. Li, P.; Zhao, C.; Zhang, Y.; Wang, X.; Wang, X.; Wang, J.; Wang, F.; Bi, Y. Calcium Alleviates Cadmium-Induced Inhibition on Root Growth by Maintaining Auxin Homeostasis in Arabidopsis Seedlings. *Protoplasma* **2016**, *253*, 185–200. [CrossRef]
49. Chen, R.; Zhang, C.; Zhao, Y.; Huang, Y.; Liu, Z. Foliar Application with Nano-Silicon Reduced Cadmium Accumulation in Grains by Inhibiting Cadmium Translocation in Rice Plants. *Environ. Sci. Pollut. Res.* **2018**, *25*, 2361–2368. [CrossRef]
50. Yamaji, N.; Ma, J.F. Node-Controlled Allocation of Mineral Elements in Poaceae. *Curr. Opin. Plant Biol.* **2017**, *39*, 18–24. [CrossRef]
51. Yamaji, N.; Xia, J.; Mitani-Ueno, N.; Yokosho, K.; Feng Ma, J. Preferential Delivery of Zinc to Developing Tissues in Rice Is Mediated by P-Type Heavy Metal ATPase OsHMA2. *Plant Physiol.* **2013**, *162*, 927–939. [CrossRef] [PubMed]
52. Finkemeier, I.; König, A.-C.; Heard, W.; Nunes-Nesi, A.; Pham, P.A.; Leister, D.; Fernie, A.R.; Sweetlove, L.J. Transcriptomic Analysis of the Role of Carboxylic Acids in Metabolite Signaling in Arabidopsis Leaves. *Plant Physiol.* **2013**, *162*, 239–253. [CrossRef] [PubMed]
53. Ostaszewska, M.; Juszczuk, I.M.; Kołodziejek, I.; Rychter, A.M. Long-Term Sulphur Starvation of *Arabidopsis thaliana* Modifies Mitochondrial Ultrastructure and Activity and Changes Tissue Energy and Redox Status. *J. Plant Physiol.* **2014**, *171*, 549–558. [CrossRef]
54. Vigani, G.; Briat, J.-F. Impairment of Respiratory Chain under Nutrient Deficiency in Plants: Does It Play a Role in the Regulation of Iron and Sulfur Responsive Genes? *Front. Plant Sci.* **2015**, *6*, 1185. [CrossRef]
55. Coppa, E.; Celletti, S.; Pii, Y.; Mimmo, T.; Cesco, S.; Astolfi, S. Revisiting Fe/S Interplay in Tomato: A Split-Root Approach to Study the Systemic and Local Responses. *Plant Sci.* **2018**, *276*, 134–142. [CrossRef]
56. McKay Fletcher, D.M.; Shaw, R.; Sánchez-Rodríguez, A.R.; Daly, K.R.; van Veelen, A.; Jones, D.L.; Roose, T. Quantifying Citrate-Enhanced Phosphate Root Uptake Using Microdialysis. *Plant Soil.* **2021**, *461*, 69–89. [CrossRef] [PubMed]
57. Tahjib-Ul-Arif, M.; Zahan, M.I.; Karim, M.M.; Imran, S.; Hunter, C.T.; Islam, M.S.; Mia, M.A.; Hannan, M.A.; Rhaman, M.S.; Hossain, M.A.; et al. Citric Acid-Mediated Abiotic Stress Tolerance in Plants. *Int. J. Mol. Sci.* **2021**, *22*, 7235. [CrossRef]
58. Liu, Y.; Zhang, C.; Zhao, Y.; Sun, S.; Liu, Z. Effects of Growing Seasons and Genotypes on the Accumulation of Cadmium and Mineral Nutrients in Rice Grown in Cadmium Contaminated Soil. *Sci. Total Environ.* **2017**, *579*, 1282–1288. [CrossRef]
59. Zhao, P.; Huang, P.; Yan, X.; Chukwuma, A.; Yang, S.; Yang, Z.; Li, H.; Yang, W. Inhibitory Effect of Exogenous Mineral Elements (Si, P, Zn, Ca, Mn, Se, Fe, S) on Rice Cd Accumulation and Soil Cd Bioavailability in Cd-Contaminated Farmlands: A Meta-Analysis. *Chemosphere* **2023**, *343*, 140282. [CrossRef]
60. Tang, L.; Mao, B.; Li, Y.; Lv, Q.; Zhang, L.; Chen, C.; He, H.; Wang, W.; Zeng, X.; Shao, Y.; et al. Knockout of OsNramp5 Using the CRISPR/Cas9 System Produces Low Cd-Accumulating Indica Rice without Compromising Yield. *Sci. Rep.* **2017**, *7*, 14438. [CrossRef]

toxics

MDPI

Article

Foliar Transpiration Inhibitor Reduces Cd Accumulation in Rice Grain: The Potential Effect of the Endophytic Bacterial Community

Ge Lei [1,2], Huijuan Song [1,*], Ziwen Gan [1], Yunchou Yang [3] and Anwei Chen [1]

[1] Department of Environment & Ecology, Hunan Agricultural University, Changsha 410128, China; lgdaniel2022@163.com (G.L.); 13807462800@163.com (Z.G.); anweihuanjing@126.com (A.C.)
[2] Department of Civil and Environmental Engineering, The Hong Kong Polytechnic University, Hong Kong, China
[3] Agricultural and Rural Affairs Bureau of Dao County Hunan Province, Yongzhou 425300, China; dxnyzh123@163.com
* Correspondence: songhj@hunau.edu.cn; Tel.: +86-0731-84617803

Abstract

Excess Cd in soils can be accumulated in rice, presenting a serious human health risk. The effect of foliar transpiration inhibitors (TIs) on the Cd content and the endophytic bacterial community in rice plants was unclear. We evaluated the key part of the rice plant to control the Cd translocation and the profile of the endophytic bacterium structure after spraying with foliar reagents; some possible typical endophytes were induced by the TIs to inhibit the Cd translocation in the rice plant. The rice plants in three sites with different available Cd content were sprayed with foliar TIs. We assessed the Cd, N, P, K and water-soluble saccharide (WSS) in different parts of the rice plant and the endophytic bacteria community in the stem. Foliar application of TIs reduced Cd translocation factor (TF_{Cd}) by ~20% from the root to the grain compared with that of CK. The TI can increase the adsorptive site concentration of stem nodes from 5.10 to 6.83 mmol/g. The diversity of the endophytic bacteria community was enhanced after application of TI, and the Shannon index increased from 3.29 to 3.92. The endophytic bacterial community induced by TI showed higher potentiality on the biofilm and stress-tolerant and metal-transport functions than that of CK, respectively. The relative abundances of *Burkholderiaceae* and *Bacterium_g_Anaeromyxobacter* were significantly negatively correlated ($p < 0.05$), with TF_{Cd} and positively correlated ($p < 0.05$), with water-solution saccharide content, simultaneously. The TI enhanced the endophytic diversity and amount. A high abundance of special endophytic bacteria induced by TI might decrease the TF_{Cd}.

Keywords: paddy field; Cd; endophytic bacteria; translocation factor; adsorptive site

1. Introduction

China faces huge challenges in soil pollution, according to the first national soil pollution survey [1]. Heavy metal is the main source of soil pollution, accounting for 82.4%, where cadmium (Cd) is the primary, accounting for 7%. It is highly mobile in soils and can be transferred through the food chain, causing damage to human health and ecosystems [2,3]. Cd tends to accumulate in rice plants, the main food crop in China [4–6]. Excessive Cd consumption can cause "itai-itai disease", as exemplified by the eating of Cd-contaminated rice in Japan in the 1960s. Therefore, how to effectively reduce the content of Cd in rice grain has become an urgent matter for the Chinese.

Foliar spray reagents are becoming a more and more important and effective technology to reduce the Cd in rice grain. The foliar reagents include six kinds [1,7]: (1) the ion antagonistic effect, such as Mn, Zn, Fe, and Ca; (2) the deposition effect, such as Si and Se; (3) the complexation effect, such as Glutathione and Cystathionine; (4) the transpiration inhibition effect, such as humic acid and salicylic acid; (5) the antioxidation effect, such as anthocyanin; and (6) the ion-channel inhibition effect, such as La and Gd. The ion antagonistic effect and transpiration inhibition were investigated relatively deeply in reducing the Cd accumulation in grain and then were applied widely in practical rice production.

The uptake and translocation of Cd in the rice plant are affected significantly by transpiration because the Cd ion is easily migrated in solution. Transpiration inhibitors (TIs) control plant transpiration and are often used to improve transplant survival. Their main components are humic acid and a lot of nutritional elements such as nitrogen, phosphorus, and potassium. The TI decreases the stomatal conductance of the leaf at high temperature periods and then reduces the Cd translocation through the low-transpiration effect. It was reported that high temperature increased Cd accumulation in rice grains through enhancing the transpiration effect [8,9]. And high-Cd-accumulation crop species are usually positively correlated with a strong transpiration effect. Therefore, the foliar application of TIs can reduce Cd uptake and accumulation in rice grain in the flowering and filling stages.

Much effort has been made to understand the mechanisms of reducing Cd accumulation in crop plants after the application of foliar reagents; however, little research has focused on the effects on the endophytic bacterial community structure. In fact, foliar reagents (containing plant hormones and other organic and inorganic compounds) can significantly influence both the activity and composition of endophytic microbial communities. These foliar reagents alter nutritional or metabolic profiles, which may in turn affect endophytic microbial activity, ultimately leading to the suppression of heavy metal translocation from soil to grains [10]. Studies have shown that foliar application of SiO_2; nanoparticles can alter the soil microbial community composition in the rhizosphere of Pakchoi (*Brassica chinensis* L.) grown in contaminated mine soil [11]. It is implied that the foliar reagent can act not only on the endophytes directly but also on the bacteria indirectly. Furthermore, the endophyte can adsorb the metal and restrict its mobilization. It is reported that the fusaric acid from a mangrove endophyte can capture the metal as Cd-, Cu-complexes [12].

The mechanisms by which endophytic microbes affect Cd translocation in rice plants primarily include adsorption, biosequestration, and the regulation of gene expression. The strong metal-binding ability of the endophytic bacterium induced the low metal translocation efficiency in the plant [13]. Endophytic *Falciphora oryzae* can sequester Cd in its vacuoles and chlamydospores, thereby enhancing Cd tolerance in rice and decreasing the translocation efficiency to shoots [14]. Alternatively, endophytic microbes can regulate the expression of some important channel proteins to adjust the uptake of essential nutrients along with mediating the uptake and translocation of heavy metals as well [15]. After the *Stenotrophomonas maltophilia* R5-5 was inoculated in rice plants, the expression of OsNramp5 and OsHMA2 was down-regulated in the root [16], and the Cd accumulation in rice grain was reduced. Typically, the Cd translocation process influenced by endophytes in the plant tends to involve more than one mechanism. However, the effect of foliar spray reagents on the endophytic community structure and the relation with Cd translocation factor (*TF*) in rice plants is unclear.

In this study, foliar reagents of TI were sprayed on rice at the flowering and filling stages, and the Cd concentration, *TF* and the profile of endophytic bacterial community within the plants were measured. The purpose of this experiment is to explore (1) the key part of rice plant to control the Cd translocation; (2) the profile of the endophytic bacterium structure after spraying with foliar reagents; and (3) some possible typical endophytes inhibiting the Cd translation in rice plant. Our investigation might provide a new viewpoint to understand the reducing of Cd accumulation in the rice grain after the foliar reagents, which could have a potential impact on the safety of rice production.

2. Materials and Methods

2.1. Experiment Site and Rice Cultivar

Field trials were conducted in Changfeng Village, Liuyang City, Hunan province, China. The soil Cd concentration at the site exceeded the national soil environmental quality standard (0.3 mg kg^{-1}). The basic physical–chemical properties in different sites are shown in Table 1. A randomized block design with three replicates was employed. The plot area was 20.0 m^2, with 5.0 m length and 4.0 m width.

Table 1. Physical–chemical properties of soil in paddy field.

Physical-Chemical Indexes	Site A	Site B	Site C
pH	6.3	7.1	7.5
Olsen-P/mg kg^{-1}	22.44	34.87	42.25
Olsen-K/mg kg^{-1}	62.25	63.54	61.75
Alkaline-N/mg kg^{-1}	80.5	78.6	76.8
Total-N/g kg^{-1}	2.57	2.56	2.87
Total Cd/mg kg^{-1}	1.2	0.85	0.54
Available Cd/mg kg^{-1}	0.21	0.15	0.12
Organic Matter/g kg^{-1}	45.95	42.52	38.54

The seed of rice (*Oryza sativa* L. cv. Zhongzao 39) was supplied by the China Rice Research Institute. Seeds were planted in March and seedlings were transplanted in late April; its whole growth period is 112 days. Field managements were as same as those used in local production. Weeds were controlled by chemical herbicide treatment.

2.2. The Components of TI and Application

The TI with humic acid purchased from the Sichuan Guoguang Agricultural Chemical Co., Ltd. (Jianyang, China) The humic acid content was 40 g/L and the nutrition element of N, P_2O_5, K_2O was 40 g/L, 80 g/L, 150 g/L, respectively. This TI solution was diluted 1000 times with water to foliar application.

2.3. Experimental Design

Beginning in mid-July 2019, the TI solution was first sprayed onto rice plants at the heading stage and then applied once per week until the grouting stage. In total, the plants received three foliar applications during the entire growth period. These foliar reagents were sprayed onto the leaves with a handheld sprayer. Control plants (CKs) were sprayed with water, and every 20 m^2 plot was sprayed with 2.0 L foliar reagents/water. At the maturity stage, three plants with roots and topsoil (0–20 cm) were collected by a shovel from the center of each plot. The roots were rinsed with tap water to remove soil. Subsequently, the samples were transported to the laboratory and washed three times with deionized water. The harvested plants were then separated into grains, stems, and roots. These grains and stems were dried for 6h at 70 °C, and the roots were stored in a constant temperature refrigerator at −70 °C.

2.4. Determination and Analysis

2.4.1. The Metal Concentration

The concentrations of Cd in plant tissues were measured with ICP (ICPMA 8300, Perkin-Elmer, Shelton, CT, USA). A 2.0 g aliquot of powdered sample was weighed into a 50 mL dry glass digestion tube. Then, 10 mL of concentrated nitric acid was added, and the mixture was left to stand overnight. The digestion tube was placed into the digestion furnace to accelerate digestion. The temperature of the digestion solution was raised to 70 °C in 15 min, kept for 30 min, and then heated up to 90 °C in 30 min; finally, it was maintained at 120 °C for 2 h until the funnel digestion became clear. After cooling to room temperature, the digest was diluted with ultrapure water to 40 mL and filtered through Whatman filter paper, and then measured by inductively coupled plasma–optical emission spectrometry (ICP-OES, Optima 5300, Perkin-Elmer, Shelton, CT, USA).

For the available Cd of soil, the soil sample was extracted with 0.01 M $CaCl_2$ solution using a soil/solution ratio of 1:5 (m:m), shaken for 2 h, and then filtered. The extract was analyzed with ICP-OES.

2.4.2. The Water-Soluble Saccharide (WSS) Content

Amounts of 0.5 g fresh stem node samples were ground in a pre-cooled mortar and subjected to extraction with a small amount of quartz sand and 5% trichloroacetic acid (TCA) solution. The sample was then centrifuged at 4000× *g* for 10 min at 4 °C, and the supernatant was collected for the measurement of the WSS content using the Bradford method [17].

2.4.3. The Nutrition for Endophytic Microorganism in the Rice Stem

We weighed 0.5000 g of ground rice stems and added 25 mL of a 2% acetic acid solution at 0.5 mol/L. This was shaken for extraction for 30 min at 25 °C and 200 rpm. Then, it was centrifuged at 10,000× *g* for 15 min, the supernatant was filtered through a 0.45 μm membrane, and the resulting filtrate was used for measurement. The soluble nitrogen (SN), soluble phosphorus (SP), and soluble potassium (SK) levels were determined using potassium persulfate oxidation UV spectrophotometry, molybdenum antimony colorimetric method and inductively coupled plasma optical emission spectroscopy (ICP-OES), respectively [18].

2.4.4. Adsorptive Site Concentration Measured with Potentiometric Titration

Potentiometric titrations were performed with the Metrohm 905 Titrando system (905 Titrando, Metrohm, Herisau, Switzerland), coupled with a GK2401C combination glass electrode. The glass electrode was calibrated using buffers at pH of 4.01, 7.00 and 10.00 before each titration. The concentration of the standard HCl and NaOH solutions were tested before the titration.

The fresh stem nodes were washed with distilled water and ethanol three times and then freeze-drying to constant weight. The 0.50 g dried stem nodes were ground to powders using liquid N_2 in a mortar. The powders were placed in a beaker with 70 mL of ultrapure water (18 MΩ·cm, 25 °C) with 0.001 M NaCl, under a N_2 atmosphere at 25 °C and then titrated with 0.1 M NaOH and 0.5 M HCl solutions. A known amount of HCl was added at the beginning of the experiment to lower the pH to approximately 2.5. The composites were equilibrated for 40 min and then titrated to pH 10 with NaOH. The background value was titrated with the same treatment using deionized water. Subsequently, a non-electrostatic model was used to fit the potentiometric titration data. The pK_a (type of functional group), total adsorption sites, and site concentration were calculated using FITEQL4.

2.5. Bacterial Richness and Diversity

The stem sections from three different locations at each level of contamination were combined and thoroughly mixed. The total microbial DNA was then extracted from each sample using the Soil Master DNA Extraction kit (Epicentre Biotechnologies, Madison, WI, USA) following the manufacturer's instructions. The V4-V5 region of the 16S ribosomal RNA (rRNA) gene was amplified using the universal primers 515F (GTGCCAGCMGC-CGCGG) and 907R (CCGTCAATTCMTTTRAGTTT) (i.e., 515907).

The oligonucleotide sequence barcode was fused to the forward primer. The PCR reaction mixture (20 μL) consisted of 4 μL of 5 FastPfu reaction buffer (TransGen Biotech, Beijing, China), 2 μL of a dNTP mixture (2.5 mM), 0.4 μL of each primer (5 mol/L), 0.4 μL of FastPfu polymerase, 10 ng of template DNA, and H_2O to reach the final volume. PCR thermal cycling was performed with an initial denaturation at 95 °C for 5 min, followed by 25 cycles of denaturation at 95 °C for 30 s, annealing at 55 °C for 30 s, and extension at 72 °C for 30 s, with a final extension period of 5 min at 72 °C. Following PCR amplification, the products were examined on a 2% (*w*/*v*) agarose gel and purified using the Maxiprep DNA Gel Extraction Kit (Axygen Biosciences, Union City, CA, USA). The purified amplicons were quantified using QuantiFluor-ST (Promega, Fitchburg, WI, USA) and then sequenced in paired-end fashion on an Illumina MiSeq platform at Majorbio Bio-Pharm Technology Co., Ltd. (Shanghai, China), following standard protocols.

The resulting sequences were clustered into operational taxonomic units (OTUs) using a 97% sequence similarity threshold. Taxonomic ranks were assigned to representative OTU sequences using the naive Bayesian Classifier (v.2.2) of the Ribosomal Database Project (RDP). Bacterial community diversity and composition were subsequently analyzed based on the OTU assignments and taxonomic classifications.

2.6. Bioinformatics and Statistical Analysis

Raw pyrosequencing data were de-multiplexed and quality-filtered using the Trimmomatic tool in the way Lohse et al. [19] described. Overlapping reads were then merged into single long reads with the FLASH software tool (https://www.cbcb.umd.edu/, accessed on 27 August 2025) [20]. Qualified sequences were then clustered into OTUs at a 97% similarity cutoff using Usearch v7.1 (http://qiime.org/, accessed on 15 November 2019). The phylogenetic affiliation of each 16S rDNA sequence was analyzed with the RDP Classifier v2.2 (http://sourceforge.net/projects/rdp-classifier/, accessed on 20 November 2019), using a confidence threshold of 0.7 and the reference database Silva (Release 115, http://www.arb-silva.de, accessed on 23 November 2019). The heatmap figures were produced using package 'gplots' in the R (v3.1.1) software (http://www.Rproject.org/, accessed on 29 November 2019).

Gene Abundance Quantification. The abundance of each gene in the non-redundant catalog within each sample was quantified by mapping the high-quality, host-filtered reads back to the catalog using Bowtie2 (v2.4.4). The number of reads mapped to each gene was counted. The abundance of a given gene was calculated and normalized as reads per kilobase per million mapped reads (RPKM) to account for gene length and sequencing depth variations between samples. The formula is:

$$\text{RPKM} = (\text{Number of reads mapped to the gene} \times 10^9)/(\text{Total number of mapped reads} \times \text{Gene length in bases})$$

This normalized value represents the relative abundance of a specific gene in the microbial community.

Functional Annotation. The non-redundant protein sequences were aligned against the Kyoto Encyclopedia of Genes and Genomes (KEGG) database (Release 106.0) using

Diamond (v2.0.15) with a blastp search (e-value $\leq 10^{-5}$). Genes were assigned to KEGG Orthology (KO) entries based on the best hit.

Identification and Abundance Profiling of Metal-Transport-Related Enzymes. KO entries associated with metal transport functions were identified by querying the KEGG BRITE hierarchies and pathway maps (e.g., ko02010 for ABC transporters, ko01100 for metabolic pathways) for relevant terms.

The relative abundance of a specific metal-transport-related enzyme in a sample was represented by the sum of the RPKM values of all non-redundant genes annotated to the corresponding KO entry. This value serves as a proxy for the functional potential of that enzyme within the endophytic bacterial community.

The overall functional profile for metal transport was constructed by aggregating the abundances of all KO entries belonging to this category.

The translocation factor (*TF*) of Cd from A part to B part of rice was defined as $TF_{A\text{–}B}$. These *TF* values were estimated as follows:

$$TF_{root\text{-}rice} = \frac{Cd_{rice}}{Cd_{root}}$$

$$TF_{stem\text{-}rice} = \frac{Cd_{rice}}{Cd_{stem\ node}}$$

$$TF_{root\text{-}stem} = \frac{Cd_{stem\ node}}{Cd_{root}}$$

where Cd_{rice}, Cd_{root} and $Cd_{stem\ node}$ are the concentrations (mg/kg, dry weight) of Cd in rice grain, roots, and stem nodes, respectively.

2.7. Statistical Analyses

The statistical analysis of the obtained results using the SPSS 19.0. Every site included two treatments: foliar TI spray treatment and control. Every treatment included three samples as three replicates. Due to the small sample size, although we conducted a normality test, its effectiveness was insufficient. Considering that this indicator is usually considered to be approximately normally distributed, we used the Student's t-test to examine statistical significance of differences in different groups. Mean $\pm$ SE (standard error) was calculated using three replicates for each treatment.

The correlation between relative abundance of endophytic bacterial species and physical parameters was determined using Spearman analysis. Prior to correlation analysis, the normality of the data distribution was assessed using the Shapiro–Wilk test.

3. Result

3.1. Effects of TI on Cd Concentration in Different Organs of Rice Plants

The Cd contents in the grains, stem nodes, and roots of rice from paddy fields at different sites are shown in Figure 1a–c. The Cd content followed the order root > stem > grain. In rice grains, TI controlled the concentration of Cd in rice below the national safety standard of 0.2 mg/kg at site B and site C (Figure 1a). The available Cd concentration in sites A, B and C were 0.21 mg/kg, 0.15 mg/kg and 0.12 mg/kg, respectively. In the stem nodes, the Cd concentration under the three treatments all decreased with decreases in the available Cd concentration in the soil (Figure 1b). The TI treatment resulted in relatively higher Cd concentrations in the stem nodes compared to the CK. These results indicate that the TI treatment promoted the accumulation of Cd in the stem nodes. The stem nodes of rice represent an important pool for Cd accumulation, which may regulate the translocation of Cd to the leaves and grains.

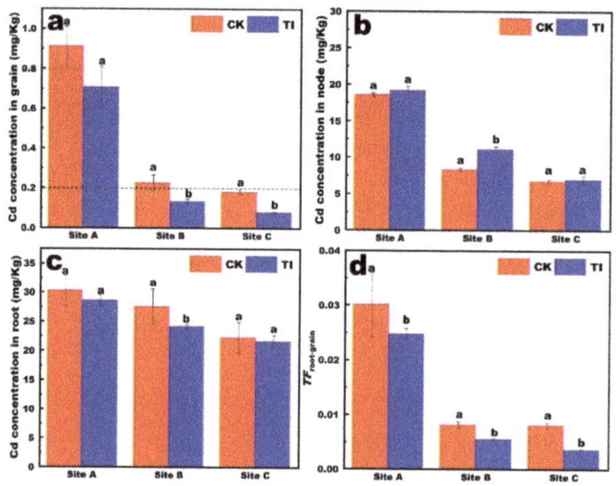

Figure 1. Effect of TI on Cd concentrations in rice grains (**a**); stem nodes (**b**); roots (**c**) and Cd translocation factor from root to rice grain (**d**) of rice plants grown in Cd-contaminated field with different Cd concentration (The letters a, b indicate statistically significant differences between treatments by Student's *t*-test. Mean ± SE (standard error) was calculated using three replicates for each treatment).

In the roots, Cd concentration reduced with the decrease in the available Cd concentration in the soil, in the TI treatments (Figure 1c). Generally, foliar application was performed with TI-sprayed aboveground parts, regulating the Cd concentration in aboveground stem and rice grain, and with no significant change in the underground root. The Cd concentration in roots was 22–30 mg/Kg, which was not a significant variation between different sites. This indicated that the Cd accumulation in roots was affected mainly by the cultivar, not the environmental Cd stress. However, the $TF_{root\text{-}rice}$ decreased significantly with the decrease in the available Cd concentration in soil. Furthermore, the $TF_{root\text{-}rice}$ with TI treatment was markedly reduced in site B and site C compared with site A (Figure 1d). These results revealed that the $TF_{root\text{-}rice}$ was affected by not only the foliar reagents, but also the soil properties.

To further understand the *TF* variation in different treatments, the histograms of *TF* in different translocation pathways of rice plants following different treatments are shown in Figure 2. The $TF_{stem\text{-}rice}$ of 0.01–0.06 was relatively lower than $TF_{root\text{-}stem}$ of 0.2–0.7. The TI can induce a higher $TF_{root\text{-}stem}$ and a lower $TF_{stem\text{-}rice}$. It was revealed that the TI induced a higher Cd content in stem node than that of CK.

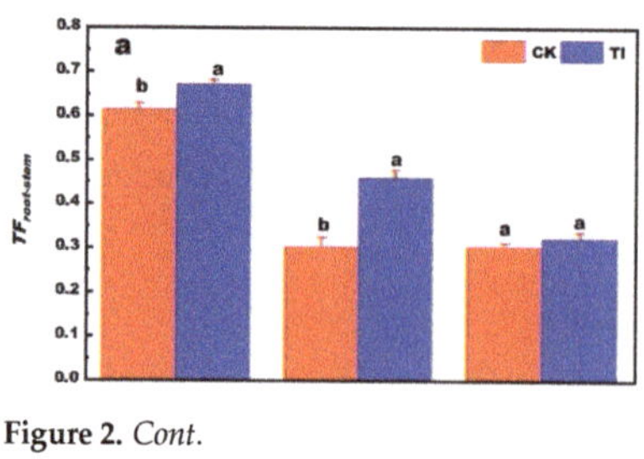

Figure 2. *Cont.*

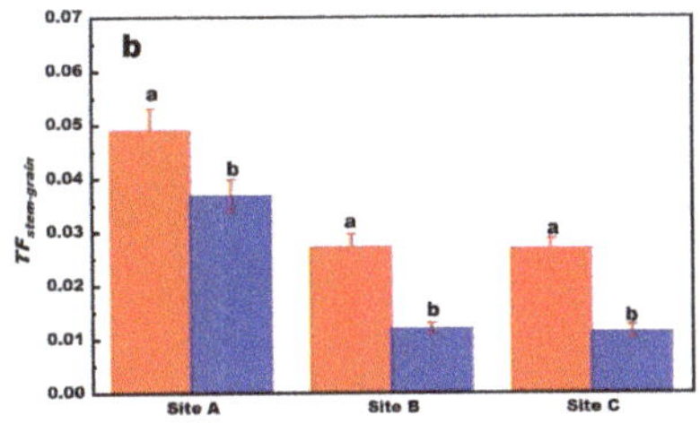

Figure 2. The translocation factors (TF) of root to stem nodes (**a**), and (**b**) stem node to rice grain, following the CK and T1 treatments in different sites (the letters a, b indicate statistically significant differences between treatments by Student's *t*-test. Mean ± SE (standard error) was calculated using three replicates for each treatment).

3.2. The WSS Content in Stem Nodes

The WSS content of stem nodes increased with decreasing the available Cd concentration in the soil, as shown in Figure 3a. Furthermore, the WSS increased at the same site with TI application than that of CK. The highest content of WSS was 0.40%, corresponding to the sample with TI treatment at site C. The relation between Cd $TF_{root\text{-}rice}$ and the WSS content is revealed in Figure 3b. The negatively linear relation was fitted with an R of 0.84. This indicated that the high WSS content corresponded to a low Cd translocation efficiency. This finding is consistent with previous reports that Cd concentration and its binding to the cell wall fraction are strongly positively correlated with lactose concentration [21]. Therefore, the WSS seems to be able to promote the Cd being adsorbed or immobilized in the organs in plants.

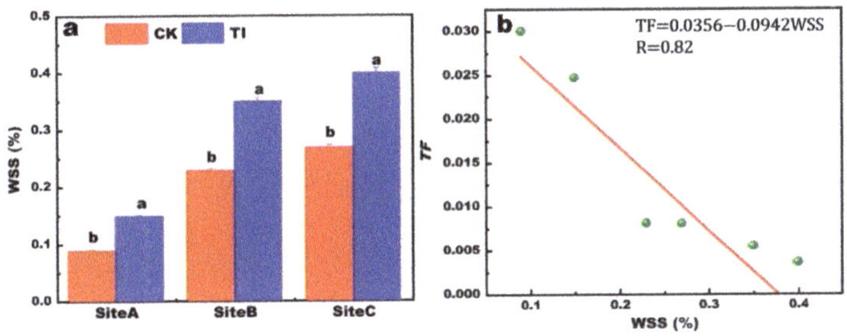

Figure 3. The content of water-soluble saccharide in stem nodes of rice plant (**a**). The relation between the $TF_{root\text{-}grain}$ and the WSS concentration (**b**). (The letters a, b indicate statistically significant differences between treatments by Student's *t*-test. Mean ± SE (standard error) was calculated using three replicates for each treatment).

3.3. The Adsorptive Site Concentration of Stem Nodes

The results of the cadmium adsorption capacity of stem nodes are shown in Table 2. The total adsorptive site concentration of TI and CK in site C were 6.83 and 5.10 mmol/g, respectively. The total adsorptive site concentration increased more in TI than that of CK, at all sites. This result corresponded to the TI inducing lower $TF_{root\text{-}rice}$ of Cd than that of CK, because of a high concentration of adsorptive sites for Cd in stem nodes. The functional groups present on biosurfaces are divided into three types [22], where pK_a values of 4–6, ~7, and 8–11 correspond to carboxyl group, phosphate group, and amino/hydroxyl group, respectively. The TI treatment increased the amino/hydroxyl groups, as shown in Table 2.

Table 2. The adsorptive site concentration obtained from stem by the potentiometric titration.

Site	Symbol	Adsorptive Site Concentration (mmol/g) ^{A}							
		Total		4 < pKa < 6		pKa ≈ 7		8 < pKa < 11	
Site A	A	4.61 ^{a}	0.08	0.96 ^{a}	0.02	1.72 ^{a}	0.04	1.93 ^{b}	0.02
	ATI	6.53 ^{c}	0.07	0.91 ^{a}	0.02	1.46 ^{b}	0.03	4.16 ^{a}	0.05
Site B	B	4.90 ^{a}	0.06	1.17 ^{a}	0.01	1.76 ^{a}	0.02	1.97 ^{b}	0.04
	BTI	6.80 ^{b}	0.08	1.02 ^{a}	0.01	1.43 ^{b}	0.02	4.35 ^{a}	0.08
Site C	C	5.80 ^{a}	0.09	0.95 ^{a}	0.02	1.82 ^{a}	0.03	3.03 ^{b}	0.05
	CTI	7.21 ^{c}	0.10	0.86 ^{a}	0.01	1.81 ^{a}	0.03	4.54 ^{a}	0.06

^{A} Average apparent acidity constant conditional to I = 0.001 mol L^{-1}. Different letters indicate statistically significant differences between treatments by Student's *t*-test.

3.4. Effects of TI on Biological Species in Rice Plants

The foliar reagents reduce the Cd content in the rice grain; this is usually attributed to the Cd translocation efficiency from the root to the rice grain being depressed. Based on the above results, the bottom node is the key organ to restrict the Cd transport upward. The mechanism might be mainly related to the high Cd adsorption capacity and chemical chelation effects. The foliar reagent can change the internal environment of the rice plant, induce the variation of the endophytic bacterial community, and then reduce the Cd transport. Therefore, the endophytic bacterial community was analyzed through a high-throughput sequencing technique to explore the potential relation between the microbial species and the Cd translocation upward in the rice plant.

High-throughput sequencing technique based on the 16S rRNA revealed the significant changes in microbial species diversity. The average total OTUs number was 622 and 716 in samples with CK (A, B, C samples) and TI (ATI, BTI, CTI samples) treatment, respectively. The Shannon index of A, B, C, ATI, BTI and CTI were 3.43 ± 0.6, 3.38 ± 0.4, 3.06 ± 0.4, 3.45 ± 0.4, 3.63 ± 0.3 and 4.68 ± 0.9, respectively. The average value of the Shannon index were 3.29 and 3.92 for CK and TI, respectively. These results are shown in Figure 4a. This indicates that the foliar TI application can improve the environment for the endophytic bacterial community and increase the microbial species diversity.

The relative abundance of main microbial species at the phylum level changed after the application of foliar reagents, as shown in Figure 4b. The Proteobacteria, Actinobacteria, Bacteroidetes, Chloroflexi were the main phylum. Their relative abundance followed the order Proteobacteria > Actinobacteria > Bacteroidetes > Chloroflexi. A heatmap depicting the endophytic bacterial community at the species level is shown in Figure 4c. The species with relative abundance higher than 1% included the *bradyrhizobium* sp_g_*Afipia, Bacterium_g_Anaeromyxobacter, Paraburkholderia kururiensis, Burkholderiaceae, Ralstonia solanacearum_g_Ralstonia, Oryza sativa Indica Group long-grained rice* g norank f *Mitochondria.*

Uncultured *bradyrhizobium* sp_g_*Afipia* decreased by 9.2% under TI treatment. Oryza sativa Indica Group long-grained *rice*_g_norank_f *Mitochondria* in TI treatment decreased by 5.48%. In contrast, *Paraburkholderia kururiensis*, unclassified_f_*Burkholderiaceae, Bacterium_g_Anaeromyxobacter* and *Ralstonia solanacearum_g_Ralstonia* increased by 1.2% to 6.25% under TI treatment. The *bradyrhizobium* sp_g_*Afipia* and *Bacterium_g_Anaeromyxobacter* are the typical N_2-fixation endophytic bacteria. *Paraburkholderia kururiensis* and unclassified_f_*Burkholderiaceae* are two types of plant growth-promoting bacteria. TI treatment can regulate the endophytic bacterial community by significantly increasing the relative abundance of plant growth-promoting bacteria and decreasing the abundance of some non-promoting bacteria.

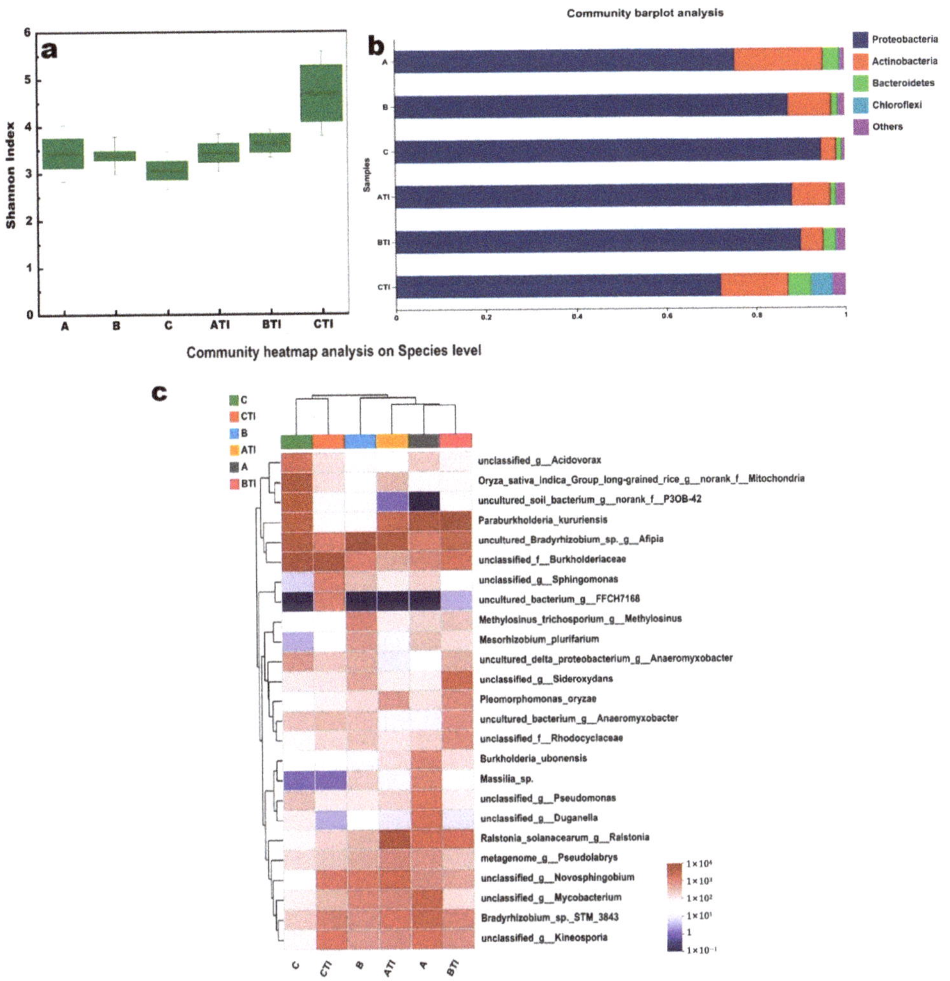

Figure 4. The Shannon index of different treatments with A, B, C, and ATI, BTI, CTI (**a**). The abundance of endophytic bacterial community on phylum level in the rice plant with different treatments (**b**). The community heatmap on species level of different treatments (**c**). The A, B, C was the CK in site A, B, C and the ATI, BTI, CTI was the samples in site A, B, C after treating with TI, respectively. Mean ± SE (standard error) was calculated using three replicates for each treatment.

3.5. The Main Predicted Function of Endophytic Bacterial Community in Different Treatments

Using the two main heavy-metal-related functions, including stress-tolerant and forms-biofilm of endophytic bacterial community induced by CK, TI treatments were predicted using the BugBase tool (version 1.0, JSON). The relative abundances of main endophytes associated with these two functions are shown in Figure 5a,b. For the forms-biofilm and stress-tolerant functions, the TI treatment showed higher function potential than that of the CK. The functions of forms-biofilm and stress-tolerant were related to the Cd-binding and biosequester abilities of the endophytic bacterial community in rice plant.

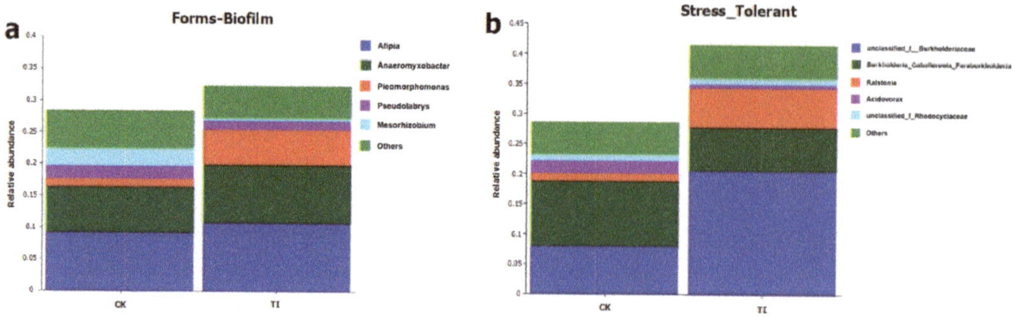

Figure 5. The relative abundance of main endophytic bacteria leading to forms-biofilms (**a**) and stress-tolerant (**b**) function in the rice plant with CK and TI treatments.

The contributions of endophytes to the forms-biofilm function mainly included *Afipia*, *Anaeromyxobacter*, *Pleomorphomonas*, *Pseudolabrys*, *Mesorhizobium*, etc. The highest relative abundance to contribute this function was *Afipia* in TI treatment, with a relative abundance of 12%.

The endophytes contribution to the stress-tolerant function mainly included unclassified_f__*Burkholderiaceae*, *Burkholderia_Caballeronia_Paraburkholderia*, *Ralstonia*, *Acidovorax*, unclassified_f_*Rhodocyclaceae*. The relative abundance of unclassified_f__*Burkholderiaceae* was as high as 23% in TI, as the most crucial contributor.

3.6. Enhanced Cd Biosequestration Potential in Endophytes

To further illustrate the potential ability of endophytes to biosequester Cd, the abundance of the metal-transport-relative enzyme in endophytes induced by different treatments is presented in Table 3. The abundance of P-type $Ca^{2+}/Zn^{2+}/Mg^{2+}/Mn^{2+}/Cu^{2+}$ transporter and Cd^{2+}-exporting ATPase for endophytes induced by TI treatment was higher than that by CK. The Cd^{2+} can be transported into endophytes' cells through these P-type $Ca^{2+}/Zn^{2+}/Mg^{2+}/Mn^{2+}/Cu^{2+}$ transporters [23]. Furthermore, the high abundance of Cd^{2+}-exporting ATPase indicates that endophytes in the TI treatment possess an enhanced capacity for Cd detoxification. It was demonstrated that the endophytes in rice plants reduced by TI had the highest Cd-sequestering ability compared to that by CK. And this result corresponded to the endophytes with high stress-tolerant function induced by TI treatment.

Table 3. The abundance of the metal-transport-relative enzyme in endophytic bacterial community in rice plant with different treatment.

Enzyme	CK	TI
P-type Ca^{2+} transporter	5774	5921
P-type Zn^{2+} transporter	11,380	13,344
P-type Mg^{2+} transporter	1315	1435
Cd^{2+}-exporting ATPase	11,380	13,344
ABC-type Mn^{2+} transporter	115	181
P-type Cu^{2+} transporter	27,352	28,362

3.7. Identification of Typical Endophytes and Their Environmental Relevance

In this study, nutrients (N, P, K), WSS and Cd, and $TF_{root\text{-}grain}$ in rice stem were taken into consideration to evaluate the relative contributions to endophytic bacteria at the species level. The detailed correlation between the bacterial species and the environmental parameters are shown in Figure 6. A species-level heatmap split the six parameters into two groups at the first level. One was composed of Cd, $TF_{root\text{-}grain}$ and K, N, while the other was composed of WSS, P.

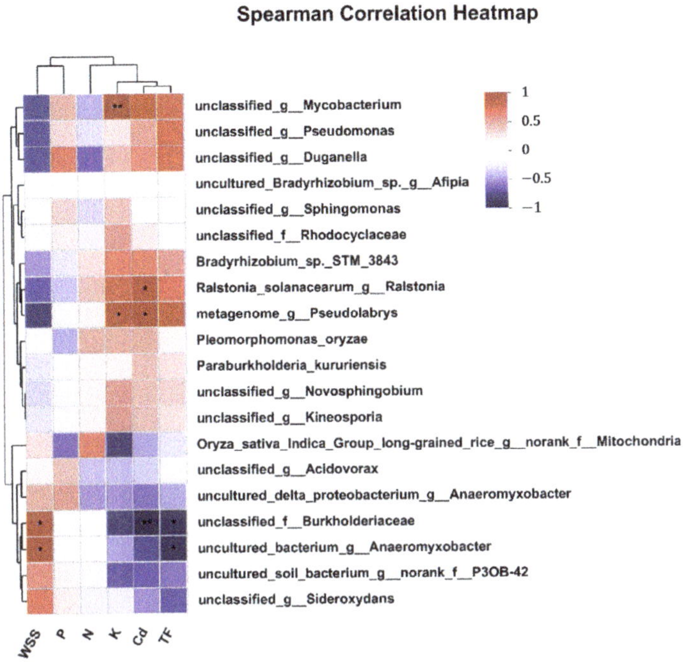

Figure 6. The heatmap of the correlation between relative abundance of endophytic bacterial species and physical parameters. WSS, P, N and K refer to the content of water-soluble saccharide, SP, SN and SK in the rice stem, respectively. And Cd, TF refer to the content of Cd in grain and $TF_{root\text{-}grain}$, respectively (* $0.01 < p \leq 0.05$, ** $0.001 < p \leq 0.01$).

Notably, unclassified_g_*Mycobacterium* was positively correlated with Cd (R = 0.77, $0.001 < p \leq 0.01$). Metagenome_g_*Pseudolabrys* was positively correlated with Cd ($R = 0.89$, $0.01 < p \leq 0.05$) and K ($R = 0.83$, $0.01 < p \leq 0.05$). *Burkholderiaceae* was negatively correlated with Cd ($R = -0.94$, $0.001 < p \leq 0.01$), $TF_{root\text{-}grain}$ ($R = -0.88$, $0.01 < p \leq 0.05$), while positively corresponded to WSS ($R = 0.89$, $0.01 < p \leq 0.05$). Uncultured_bacterium_g_*Anaeromyxobacter* was negatively correlated with $TF_{root\text{-}grain}$ ($R = -0.86, 0.01 < p \leq 0.05$), while positively corresponded to WSS ($R = 0.86, 0.01 < p \leq 0.05$).

The relationship between the WSS content and the microbial species was investigated as well. Two microbes were positively correlated with WSS ($p \leq 0.05$), i.e., *Burkholderiace* and *Bacterium_g_Anaeromyxobacter*. This suggests that an increase in the relative abundance of these specific endophytic bacteria may contribute to a decrease in the $TF_{root\text{-}rice}$. Figure 7 presents the detailed MaAslin analysis of the relationships between the two microbes and both $TF_{root\text{-}rice}$ and WSS content. The high relative abundance of *Burkholderiace* and *Bacterium_g_Anaeromyxobacter* was revealed in the rice plant with different treatments, respectively.

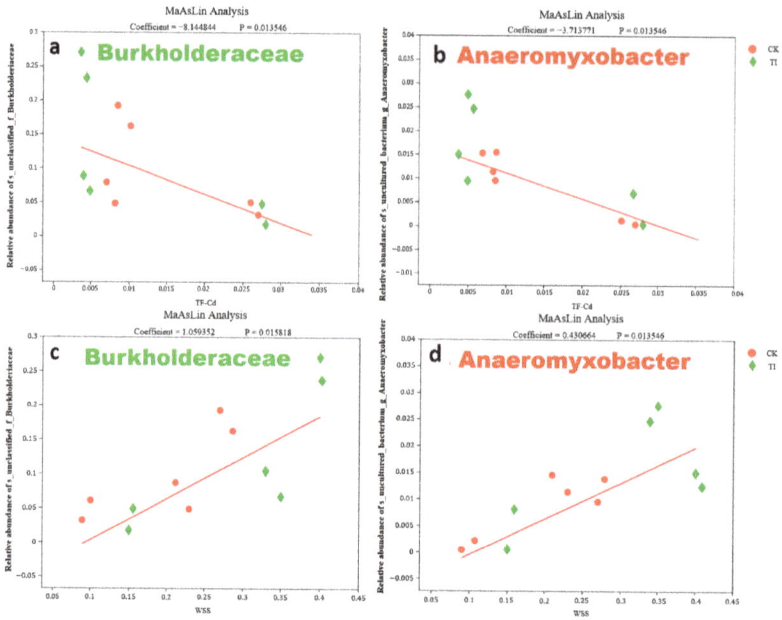

Figure 7. The relation between the relative abundance of *Burkholderiace, Bacterium_g_Anaeromyxobacter* and Cd translocation factor from root to grain (**a**,**b**), WSS content (**c**,**d**) in rice plant, respectively.

4. Discussion

4.1. The TI Enhancing the Cd Accumulation in the Stem Node

TI treatment resulted in a higher $TF_{root\text{-}stem}$ and a lower $TF_{stem\text{-}rice}$, suggesting that Cd accumulation was enhanced in the stem nodes rather than in the rice grain. The foliar reagents reduce all the translocation efficiency of Cd in every part, in that the change of Cd accumulation in rice is related to Cd transport genes. The OsHMA3 [24], NRAMP5 [25], and OsHMA2 [26] have been found to play a role in Cd absorption/transport in rice, and the gene expression of these transporters is usually related to the application of foliar reagents [27]. Prior to the application of the foliar reagent, a substantial amount of Cd had already accumulated in the stem nodes. Subsequently, the transportation of Cd was depressed after the foliar application with TI at the heading stage of rice plant growth. Therefore, TI induced the relatively lower $TF_{stem\text{-}grain}$, and the Cd was accumulated at the stem nodes.

Our results revealed that the stem nodes exhibited a higher Cd accumulation capacity after spraying with TI, due to their higher concentration of adsorptive sites compared to the CK. The adsorptive sites were mainly attributed to the cell wall, endophytes and the extracellular polymeric substances (EPS) secreted by endophytes. However, the precise relationships between specific functional groups and individual components could not be unequivocally identified in this study. Mainly, the hydroxyl group was correlated with a silicon-containing cell wall, and the phosphate and amino groups were related to the EPS [28] and endophytes [29]. Therefore, the high adsorptive site concentration might be attributed to the large amount of endophytes. Furthermore, this was supported by the observation that the concentration of WSS was also higher in TI-treated plants than in CK-treated plants.

The WSS content might be correlated with the amount of endophytes. On the measurement process, the endophytes were extracted from the rice plant as the component of WSS.

Therefore, a high number of endophytes induced by foliar TI resulted in the high adsorptive site concentration in rice plant to reduce the Cd translocation upward in rice plant.

4.2. TI Enhancing the Endophytic Diversity and Cd Uptake Capacity

TI increased the endophytic community diversity. Foliar spraying of TI changed the community structure of the colonizing bacteria and regulated the internal ecological environment of rice plant. Tian et al. [11], spraying foliar nano SiO_2 on pakchoi, found that the rhizosphere bacterial and fungal were increased. And these bacteria promoted the carbon and nitrogen cycle, which led to increasing root secretion and the production of heavy metal adsorption chelation, thereby improving the pollution-resistance effect. It was indicated that the foliar reagent could change the relative abundance of endophytic bacterial species, through altering the internal nutrition and ecological environment in plants.

TIs increase the forms-biofilm and the stress-tolerant functions of endophytes. The biofilms are usually composed of extracellular polysaccharide, protein, and lipid. The biofilms have been considered to have a high Cd-binding capacity, and with a large adsorptive site [30]. Therefore, an enhanced biofilm-forming capability likely contributes to a greater Cd-binding effect. Normally, the stress tolerance of endophytes to heavy metal partly reflects its detoxification effect, because the high heavy metal concentration can induce the high content of antioxidant enzymes, to alleviate the reactive oxygen and heavy metal damage [31]. The endophytic microbes with excellent Cd biosequestering capacity might correspond to their high detoxification effect; otherwise, the high Cd content would damage its physiological activity.

4.3. Typical Endophytes Potential Effect on Reduce Cd Translocation

Our research results indicate that among rice plants subjected to different treatments, both *Burkholderiace* and *Bacterium_g_Anaeromyxobacter* exhibit relatively high abundance. *Burkholderia* cepacian complex bacteria usually have high resistance to Cd, Cu, Zn, Co, Pb and Hg [32]. It is noted that the *Burkholderia* can colonize in the root system of plants [33] and compete with the root cells to sequester Cd, which reduces the Cd upward transport. Wang et al. [34] found that the *Burkholderia* sp. Y4 inoculation in rice plant would increase essential nutrient uptake and inhibit Cd accumulation in rice by preferential Cd-biosequester.

The *Bacterium_g_Anaeromyxobacter* can drive reductive dechlorination of chlorophenols [35]. *Anaeromyxobacter* spp. can defend against other bacteria using exoenzymes [36]. The *Bacterium_g_Anaeromyxobacter* induced increasing the Cd concentration and decreasing the Cu concentration in the roots. The endophytes can improve the Cd binding on the cell wall surface through the function groups such as -OH, $-NH_2$, $-PO_4$ and -COOH. The *Anaeromyxobacter* strain can secrete abundant EPS [35] to adsorb Cd as well. Similar research [37] showed that Cd-immobilizing's *Serratia liquefaciens* CL-1resulted in reducing Cd translocation upward in rape plant. Therefore, the large number of special endophytes might induce high adsorption effect of rice plant to Cd, and then decrease $TF_{root\text{-}rice}$ of Cd.

5. Conclusions

The foliar spray application of TIs can efficiently reduce the Cd translocation from root to rice and decrease the Cd accumulation in the brown rice in the paddy soil with different available Cd concentration. The low Cd transport from root to rice grain was mainly restricted by the nodes because of its high adsorptive site concentration. The foliar TI enhances the endophytic bacterial diversity. The relative abundance of *Bacterium_g_Anaeromyxobacter* and *Burkholderiaceae* were increased by TI treatments, respectively. Furthermore, the endophytic bacterial community induced by TIs showed greater ability under forms-biofilm and stress-tolerant functions than that by CK. It was indicated that the endophytes induced by TI revealed a greater potential for Cd-binding and bio-sequestering effect, respectively. The relative abundance of *Bacterium_g_Anaeromyxobacter* and *Burkholderiaceae* were significantly correlated with the low Cd transport upward. Our findings demonstrated these endophytic bacteria's diversity and relative abundance can be regulated by the foliar reagents. And some key endophytic microbes would regulate Cd translocation efficiency.

Author Contributions: Investigation, G.L., H.S., Z.G., Y.Y. and A.C.; Methodology, G.L. and H.S.; Writing—original draft, G.L., H.S., Z.G., Y.Y. and A.C.; Writing—review and editing, G.L. and H.S. All authors have read and agreed to the published version of the manuscript.

Funding: Key Technologies Research Development Program (2022YFD1700102), Key Projects of the Hunan Education Department (23A0165).

Data Availability Statement: The original contributions presented in this study are included in the article. Further inquiries can be directed to the corresponding author.

Conflicts of Interest: The authors have no relevant financial or non-financial interests to disclose.

References

1. Chen, R.; Zhang, C.; Zhao, Y.; Huang, Y.; Liu, Z. Foliar application with nano-silicon reduced cadmium accumulation in grains by inhibiting cadmium translocation in rice plants. *Environ. Sci. Pollut. Res.* **2018**, *25*, 2361–2368. [CrossRef]
2. Pipoyan, D.; Beglaryan, M.; Costantini, L.; Molinari, R.; Merendino, N. Risk assessment of population exposure to toxic trace elements via consumption of vegetables and fruits grown in some mining areas of Armenia. *Hum. Ecol. Risk Assess. Int. J.* **2018**, *24*, 317–330. [CrossRef]
3. Zeng, L.; Zhou, F.; Zhang, X.; Qin, J.; Li, H. Distribution of heavy metals in soils and vegetables and health risk assessment in the vicinity of three contaminated sites in Guangdong Province, China. *Hum. Ecol. Risk Assess. Int. J.* **2018**, *24*, 1901–1915. [CrossRef]
4. Cao, F.; Wang, R.; Cheng, W.; Zeng, F.; Ahmed, I.M.; Hu, X.; Zhang, G.; Wu, F. Genotypic and environmental variation in cadmium, chromium, lead and copper in rice and approaches for reducing the accumulation. *Sci. Total Environ.* **2014**, *496*, 275–281. [CrossRef] [PubMed]
5. Liu, Y.; Zhang, C.; Zhao, Y.; Sun, S.; Liu, Z. Effects of growing seasons and genotypes on the accumulation of cadmium and mineral nutrients in rice grown in cadmium contaminated soil. *Sci. Total Environ.* **2017**, *579*, 1282–1288. [CrossRef] [PubMed]
6. Wei, B.; Yang, L. A review of heavy metal contaminations in urban soils, urban road dusts and agricultural soils from China. *Microchem. J.* **2010**, *94*, 99–107. [CrossRef]
7. Han, X.; Zhang, C.; Wang, C.; Huang, Y.; Liu, Z. Gadolinium inhibits cadmium transport by blocking non-selective cation channels in rice seedlings. *Ecotoxicol. Environ. Saf.* **2019**, *179*, 160–166. [CrossRef]
8. Guo, H.; Zhu, J.; Zhou, H.; Sun, Y.; Yin, Y.; Pei, D.; Ji, R.; Wu, J.; Wang, X. Elevated CO_2 levels affects the concentrations of copper and cadmium in crops grown in soil contaminated with heavy metals under fully open-air field conditions. *Environ. Sci. Technol.* **2011**, *45*, 6997–7003. [CrossRef]
9. Liao, Y.; Chien, S.C.; Wang, M.; Shen, Y.; Hung, P.; Das, B. Effect of transpiration on Pb uptake by lettuce and on water soluble low molecular weight organic acids in rhizosphere. *Chemosphere* **2006**, *65*, 343–351. [CrossRef]
10. Means, N.E.; Kremer, R.J.; Ramsier, C. Effects of glyphosate and foliar amendments on activity of microorganisms in the soybean rhizosphere. *J. Environ. Sci. Health Part B* **2007**, *42*, 125–132. [CrossRef]
11. Tian, L.; Shen, J.; Sun, G.; Wang, B.; Ji, R.; Zhao, L. Foliar Application of SiO_2 Nanoparticles Alters Soil Metabolite Profiles and Microbial Community Composition in the Pakchoi (*Brassica chinensis* L.) Rhizosphere Grown in Contaminated Mine Soil. *Environ. Sci. Technol.* **2020**, *54*, 13137–13146. [CrossRef]
12. Pan, J.-H.; Chen, Y.; Huang, Y.-H.; Tao, Y.-W.; Wang, J.; Li, Y.; Peng, Y.; Dong, T.; Lai, X.-M.; Lin, Y.-C. Antimycobacterial activity of fusaric acid from a mangrove endophyte and its metal complexes. *Arch. Pharmacal Res.* **2011**, *34*, 1177–1181. [CrossRef]
13. Wang, Y.; Luo, H.; Peng, H.; Wang, X.; Xu, F.; Xu, H. Coprinus comatus endophytic bacteria characteristics and mechanisms for the cadmium resistance. *Environ. Sci. Pollut. Res.* **2021**, *29*, 584–593. [CrossRef] [PubMed]
14. Su, Z.-Z.; Dai, M.-D.; Zhu, J.-N.; Liu, X.-H.; Li, L.; Zhu, X.-M.; Wang, J.-Y.; Yuan, Z.-L.; Lin, F.-C. Dark septate endophyte Falciphora oryzae-assisted alleviation of cadmium in rice. *J. Hazard. Mater.* **2021**, *419*, 126435. [CrossRef] [PubMed]
15. Liu, F.; Liu, X.; Ding, C.; Wu, L. The dynamic simulation of rice growth parameters under cadmium stress with the assimilation of multi-period spectral indices and crop model. *Field Crops Res.* **2015**, *183*, 225–234. [CrossRef]
16. Zhou, J.; Li, P.; Meng, D.; Gu, Y.; Zheng, Z.; Yin, H.; Zhou, Q.; Li, J. Isolation, characterization and inoculation of Cd tolerant rice endophytes and their impacts on rice under Cd contaminated environment. *Environ. Pollut.* **2020**, *260*, 113990. [CrossRef]

17. Bradford, M.M. A rapid and sensitive method for the quantitation of microgram quantities of protein utilizing the principle of protein-dye binding. *Anal. Biochem.* **1976**, *72*, 248–254. [CrossRef]
18. Bao, S.D. *Soil and Agricultural Chemistry Analysis*, 3rd ed.; China Agriculture Press: Beijing, China, 2000.
19. Lohse, M.; Bolger, A.M.; Nagel, A.; Fernie, A.R.; Lunn, J.E.; Stitt, M.; Usadel, B. RobiNA: A user-friendly, integrated software solution for RNA-Seq-based transcriptomics. *Nucleic Acids Res.* **2012**, *40*, W622–W627. [CrossRef]
20. Mago, T.; Salzberg, S.L. FLASH: Fast length adjustment of short reads to improve genome assemblies. *Bioinformatics* **2011**, *27*, 2957–2963. [CrossRef]
21. Cheng, Y.; Bao, Y.; Chen, X.; Yao, Q.; Wang, C.; Chai, S.; Zeng, J.; Fan, X.; Kang, H.; Sha, L.; et al. Different nitrogen forms differentially affect Cd uptake and accumulation in dwarf Polish wheat (*Triticum polonicum* L.) seedlings. *J. Hazard. Mater.* **2020**, *400*, 123209. [CrossRef]
22. Fein, J.B.; Daughney, C.J.; Yee, N.; Davis, T.A. A chemical equilibrium model for metal adsorption onto bacterial surfaces. *Geochim. Cosmochim. Acta* **1997**, *61*, 3319–3328. [CrossRef]
23. Nanda, S.; Mohanty, B.; Joshi, R.K. Endophyte-Mediated Host Stress Tolerance as a Means for Crop Improvement. In *Endophytes and Secondary Metabolites*; Springer: Cham, Switzerland, 2019; pp. 677–701.
24. Miyadate, H.; Adachi, S.; Hiraizumi, A.; Tezuka, K.; Nakazawa, N.; Kawamoto, T.; Katou, K.; Kodama, I.; Sakurai, K.; Takahashi, H.; et al. OsHMA3, a P_{1B}-type of ATPase affects root-to-shoot cadmium translocation in rice by mediating efflux into vacuoles. *New Phytol.* **2011**, *189*, 190–199. [CrossRef]
25. Sasaki, A.; Yamaji, N.; Yokosho, K.; Ma, J.F. Nramp5 Is a Major Transporter Responsible for Manganese and Cadmium Uptake in Rice. *Plant Cell* **2012**, *24*, 2155–2167. [CrossRef] [PubMed]
26. Satoh-Nagasawa, N.; Mori, M.; Nakazawa, N.; Kawamoto, T.; Nagato, Y.; Sakurai, K.; Takahashi, H.; Watanabe, A.; Akagi, H. Mutations in Rice (*Oryza sativa*) Heavy Metal ATPase 2 (OsHMA2) Restrict the Translocation of Zinc and Cadmium. *Plant Cell Physiol.* **2012**, *53*, 213–224. [CrossRef] [PubMed]
27. Gao, M.; Zhou, J.; Liu, H.; Zhang, W.; Hu, Y.; Liang, J.; Zhou, J. Foliar spraying with silicon and selenium reduces cadmium uptake and mitigates cadmium toxicity in rice. *Sci. Total Environ.* **2018**, *631–632*, 1100–1108. [CrossRef] [PubMed]
28. Tourney, J.; Ngwenya, B.T.; Mosselmans, J.F.; Tetley, L.; Cowie, G.L. The effect of extracellular polymers (EPS) on the proton adsorption characteristics of the thermophile Bacillus licheniformis S-86. *Chem. Geol.* **2008**, *247*, 1–15. [CrossRef]
29. Cheng, C.; Wang, R.; Sun, L.; He, L.; Sheng, X. Cadmium-resistant and arginine decarboxylase-producing endophytic Sphingomonas sp. C40 decreases cadmium accumulation in host rice (*Oryza sativa* Cliangyou 513). *Chemosphere* **2021**, *275*, 130109. [CrossRef]
30. Song, H.; Kuang, X.; Wei, X.; Luo, S.; Zeng, Q.; Peng, L. The Effect of TiO_2 Nanoparticles Size on Cd (II) Removal by the Paddy Crusts from Waterbody. *J. Environ. Chem. Eng.* **2022**, *10*, 107883. [CrossRef]
31. Sterckeman, T.; Thomine, S. Mechanisms of cadmium accumulation in plants. *Crit. Rev. Plant Sci.* **2020**, *39*, 322–359. [CrossRef]
32. Coenye, T. The Family Burkholderiaceae. In *The Prokaryotes*; Springer: Berlin/Heidelberg, Germany, 2014. [CrossRef]
33. Chen, Z.-J.; Sheng, X.-F.; He, L.-Y.; Huang, Z.; Zhang, W.-H. Effects of root inoculation with bacteria on the growth, Cd uptake and bacterial communities associated with rape grown in Cd-contaminated soil. *J. Hazard. Mater.* **2013**, *244–245*, 709–717. [CrossRef]
34. Wang, C.; Huang, Y.; Yang, X.; Xue, W.; Zhang, X.; Zhang, Y.; Pang, J.; Liu, Y.; Liu, Z. *Burkholderia* sp. Y4 inhibits cadmium accumulation in rice by increasing essential nutrient uptake and preferentially absorbing cadmium. *Chemosphere* **2020**, *252*, 126603. [CrossRef]
35. Sanford, R.A.; Cole, J.R.; Tiedje, J.M. Characterization and description of *Anaeromyxobacter dehalogenans* gen. nov., sp. nov., an aryl-halorespiring facultative anaerobic myxobacterium. *Appl. Environ. Microbiol.* **2002**, *68*, 893–900. [CrossRef]
36. Reichenbach, H. Myxobacteria, producers of novel bioactive substances. *J. Ind. Microbiol. Biotechnol.* **2001**, *27*, 149–156. [CrossRef]
37. Han, H.; Wang, Q.; He, L.-Y.; Sheng, X.-F. Increased biomass and reduced rapeseed Cd accumulation of oilseed rape in the presence of Cd-immobilizing and polyamine-producing bacteria. *J. Hazard. Mater.* **2018**, *353*, 280–289. [CrossRef]

toxics

MDPI

Article

Effects of Malic Acid on Cadmium Uptake and Translocation and Essential Element Accumulation in Rice

Shuo Zhang [1], Yiteng Zhang [1], Guoyi Lv [1], Tianqi Liu [2], Zhongqi Liu [3], Yubo Jiang [1], Yubo Hao [1], Yang Yu [1], Wenjun Dong [1] and Chunrong Qian [1,*]

[1] Heilongjiang Academy of Agricultural Sciences, Harbin 150086, China; shuozhang994@163.com (S.Z.); yiteng_fighting@163.com (Y.Z.); lsyjtx2025@163.com (G.L.); vbojiang2007@163.com (Y.J.); yubohao2005@163.com (Y.H.); yuyanghaas@163.com (Y.Y.); dongwenjun0911@163.com (W.D.)

[2] College of Resources and Environment, Northeast Agricultural University, Harbin 150030, China; cppc161@163.com

[3] Key Laboratory of Original Agro-Environmental Pollution Prevention and Control, Agro-Environmental Protection Institute, Ministry of Agriculture and Rural Affairs, Tianjin 300191, China; liuzhongqi508@163.com

* Correspondence: qianchunrong@haas.cn

Abstract

Cadmium (Cd) contamination poses a serious threat to rice safety and productivity. This study investigated the potential of malic acid (MA), a key metabolic organic acid, to mitigate Cd toxicity and its genotype-dependent effects on cadmium uptake and essential element homeostasis in rice. Using hydroponic experiments with multiple genotypes, we found that MA application (0.5–1.5 mmol·L^{-1}) significantly reduced Cd accumulation in both roots and shoots, with the most effective reduction (up to 68.0%) achieved at 1.5 mmol·L^{-1}. Notably, genotype X24 was a low-Cd accumulator, while genotypes 20, 58, and 65 were high accumulators. Beyond Cd reduction, this study reveals the profound and genotype-specific modulation of nutrient homeostasis by MA, including consistent suppression of K and enhancement of Ca across genotypes, and highly divergent responses in Mg, Mn, Fe, and Zn accumulation. Furthermore, MA dramatically alleviated Cd-induced inhibition of root morphology, particularly in the high-Cd genotype 58, increasing root length and tip number by 42.8% and 57.8%, respectively. Our results provide novel insights into the genotype-dependent rebalancing of essential elements under MA amendment, highlighting the crucial role of genetic background in plant responses to organic acid treatments. These findings provide a mechanistic basis for developing MA-based foliar conditioners and genotype-specific strategies for managing Cd contamination in rice.

Keywords: rice; cadmium; malic acid; essential element; genotype

1. Introduction

Rice (*Oryza sativa* L.) is one of China's staple food crops. In recent years, Cd contamination in rice has become increasingly severe [1]. Annual pollution affects approximately 12 million tons of grain, resulting in economic losses estimated at CNY 20 billion [2,3].

Cd enters rice grains through multiple pathways, including atmospheric deposition, sewage irrigation, sludge application, and heavy metal-enriched fertilizers [4]. Atmospheric deposition is the primary source of Cd pollution in industrial areas [5]. Industrial and mining activities emit large amounts of Cd-containing gases and dust into the atmosphere. These pollutants enter the soil or adhere to rice plants via gravity or rainfall, subsequently being absorbed by roots and leaves and transported to grains during the filling stage.

The closer the rice fields are to Cd pollution sources, the higher the Cd content in the grains [6,7]. In some industrialized countries, atmospheric deposition contributes nearly 50% of the Cd found in rice [8]. In China, sewage irrigation represents another major Cd pollution source [9], with increasing volumes of untreated wastewater elevating Cd levels in paddies [10–12]. Although sludge improves soil quality, it can also introduce Cd and other heavy metals [13–15]. Furthermore, phosphate fertilizers—some containing up to 174 $mg \cdot kg^{-1}$ Cd—have become a significant source of soil Cd pollution after long-term application [16–19].

Cd enters rice roots via both active (carrier protein-mediated, energy-dependent symplastic transport) and passive (electrochemical potential-driven apoplastic transport) mechanisms [20]. Following uptake, a portion of Cd is immobilized in roots (e.g., bound to cell walls, organic acids, or proteins [21–23]), while the remainder is translocated via the xylem to shoots, stems, and panicles, and finally accumulates in grains during filling [20]. Under Cd stress, rice activates detoxification pathways to mitigate toxicity [21–25], though the efficiency of these mechanisms—and thus Cd accumulation capacity—varies significantly across genotypes [26]. For example, hybrid rice exhibits stronger Cd uptake and transport abilities than conventional rice [27], and differences in root absorption and xylem transport efficiency are key drivers of genotypic variation in grain Cd content [28–31]. Additionally, the phloem plays a critical role in Cd translocation to grains, with variability in phloem transport further contributing to genotypic differences in grain Cd levels [32].

Cd toxicity also disrupts various physiological processes in rice: it damages chloroplast ultrastructure, reduces photosynthetic rate, stomatal conductance [33], and transpiration [34], and accelerates leaf chlorosis. Notably, chlorophyll stability and the rate of chlorosis have been proposed as practical indicators for rapidly assessing Cd tolerance among rice varieties [35]. Beyond impairing crop productivity, Cd accumulated in rice grains enters the food chain, posing serious health risks to humans and animals [36–39]. These impacts highlight the urgency of developing effective strategies to reduce Cd accumulation in rice.

Malic acid (MA), a key small-molecule organic acid in the tricarboxylic acid (TCA) cycle, links metabolic processes across plant organelles and plays pivotal roles in plant stress adaptation [40]. Under stress conditions, plants produce and secrete organic acids to regulate rhizosphere cation exchange capacity, modulate nutrient balance, and enhance tolerance to toxic metals [41,42]. For heavy metals like Cd, organic acids can bind to Cd in the xylem, converting it into less toxic forms and alleviating translocation-related toxicity [43]. Exogenous MA has been shown to promote plant growth, increase net photosynthetic rate, reduce H_2O_2 accumulation, and improve root activity under Cd stress [44], while root-secreted MA is closely associated with Cd tolerance in rice. However, conflicting findings exist—some studies suggest root-secreted organic acids may mobilize soil Cd, increasing its bioavailability and promoting accumulation in certain plants [45,46].

Given rice's high Cd enrichment capacity among staple cereals [47–49], and the need for effective, sustainable Cd mitigation strategies, this study aimed to (1) determine the optimal MA concentration for inhibiting Cd uptake and translocation in rice seedlings; (2) elucidate the genotype-dependent effects of MA on Cd accumulation; and (3) explore how MA modulates essential element homeostasis under Cd stress. The findings are expected to provide a mechanistic basis for developing MA-based foliar conditioners and genotype-specific Cd management strategies for rice in heavy-metal-polluted regions.

2. Materials and Methods

2.1. Experimental Materials

The experimental materials are rice germplasm resources preserved in the laboratory and rice genotypes grown on a large scale in the south of China (Table 1). All chemical reagents used in this study, unless otherwise specified, were of analytical grade and purchased from Sinopharm Chemical Reagent Co., Ltd. (Shanghai, China).

Table 1. List of different rice materials.

List	Genotype
X24	Xiang zao shan 24
20	Nan xiong zao you zhan
58	Zhu zhen B
65	Zhen shan 97B

2.2. Hydroponic Culture Test

Uniform and healthy rice seeds were selected, sterilized with 5% sodium hypochlorite for 15 min, rinsed with deionized water, and placed in sterilized trays with deionized water covering the seeds. The trays were incubated at 30 °C in the dark for germination. After root and shoot emergence, seedlings were transferred to a climate chamber for light exposure. At the two-leaf stage, seedlings were moved to 8 L hydroponic boxes (100 seedlings per box) containing 1/10 Hoagland nutrient solution. This setup ensured that from the two-leaf stage onward, all seedlings were consistently maintained in the nutrient solution.

Treatments were initiated at the three-leaf stage. The experiment adopted a randomized complete block design (RCBD), with each 8 L hydroponic box serving as an independent experimental unit. All hydroponic boxes were randomly placed in the climate chamber to eliminate potential positional interference (e.g., uneven light or temperature distribution). The experiment comprised a control group (CK) and three treatment groups, each with three independent biological replicates ($n = 3$). The compositions were as follows: CK (no additions), M0 (0.3 mg·L^{-1} Cd), M0.5 (0.3 mg·L^{-1} Cd + 0.5 mmol·L^{-1} malic acid), and M1.5 (0.3 mg·L^{-1} Cd + 1.5 mmol·L^{-1} malic acid). Prior to treatment, seedlings from all groups underwent a 24 h pre-starvation in deionized water to standardize their nutritional status. Following this, the standard 1/10 Hoagland solution was applied and refreshed every two days for a 10-day treatment period. The pH of all solutions was adjusted to and maintained within the range of 5.5–6.0 throughout the experiment. The climate chamber conditions were light period 25 °C, 16 h, 40% humidity; dark period 20 °C, 8 h, 60% humidity. After treatment, rice seedling roots were immersed in 5 mmol·L^{-1} $CaCl_2$ for 15 min to remove surface Cd^{2+}, followed by three rinses with deionized water. Some seedlings were used for root scanning, while others were dried to constant weight for Cd and essential element analysis.

2.3. Determination of Root Morphology

Washed rice seedlings ($n = 20$) were scanned using a root scanner (Perfection V700 Photo, Seiko Epson Corp., Tokyo, Japan). The resulting images were analyzed with WinRHIZO software (version 2009a, Regent Instruments, Québec City, QC, Canada) [50] to measure total root length, root surface area, root volume, and root tip number.

2.4. Determination of Cd and Essential Elements

Rice root and shoot samples (0.5 g each) were digested with 7 mL HNO_3, left at room temperature for 12 h, and heated at 110 °C for 2.5 h. After cooling, 1 mL H_2O_2 was added, and heating continued for 1.5 h. The digest was evaporated to <1 mL at 170 °C, diluted, and analyzed for Cd, K, Ca, Mg, Fe, Mn, and Zn using ICP-MS (NexION®1000, PerkinElmer, Waltham, MA, USA) at Nanjing Webiolotech Testing Technology Co., Ltd. (Nanjing, China).

2.5. Data Processing and Calculation Method

Data were analyzed using IBM SPSS statistics (version 26.0). Significant differences among treatments were determined by Duncan's multiple range test at $p < 0.05$ level. All data are presented as mean ± standard deviation (SD) of three biological replicates. Figures were prepared using Origin 8.6 (version 8.6).

3. Results

3.1. Genotypic Differences in Cadmium Accumulation in Rice

Rice seedlings of genotypes X24, 20, 58, and 65 were subjected to 0.3 mg·L^{-1} Cd stress for ten days, and significant differences in Cd accumulation in shoots and roots were observed among the rice genotypes (Figure 1). Among the four genotypes, X24 was identified as a low-Cd-accumulating variety, with significantly lower Cd concentrations in both shoots and roots compared to genotypes 58 and 65 (Figure 1). Genotype 58 showed the highest shoot Cd accumulation, while genotype 65 had the highest root Cd accumulation, consistent with their classification as high-Cd accumulators. In summary, under 0.3 mg·L^{-1} Cd stress, genotypes 58 and 65 showed slightly higher Cd accumulation capacity than genotype 20, which in turn accumulated more Cd than genotype X24.

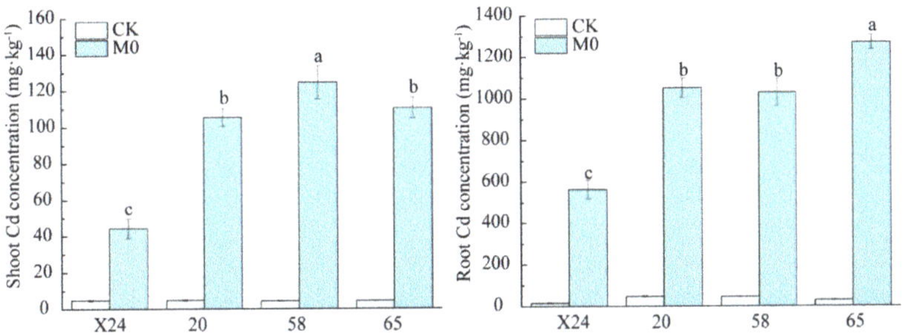

Figure 1. Cadmium concentration in rice root and shoot under 0.3 mg·L^{-1} Cd exposure. CK: Untreated control group (no additions); M0: 0.3 mg·L^{-1} Cd (sole Cd stress treatment). Different letters indicate a significant difference between different genotypes at 5% level.

3.2. Effects of Cd on Essential Element Accumulation in Rice Genotypes

To explore how Cd stress interacts with genotypic differences to affect nutrient uptake, we analyzed the accumulation of essential elements (K, Ca, Mg, Fe, Mn, Zn) in roots and shoots of the four genotypes under Cd-free and Cd-stressed conditions.

Under the Cd-free condition, significant genotypic differences in essential element accumulation were observed among the four rice genotypes. In conditions without Cd stress, root Mg, K, and Zn accumulation was notably more prominent in genotype 65, while X24 demonstrated higher levels of Ca and Mn (Figure 2, Table A1). In the shoots, X24 maintained elevated Mg, Ca, and Mn content; genotype 58 had the highest K accumulation; and genotype 65 showed the greatest Zn accumulation. These results clearly reveal the distinct patterns of elemental accumulation among different genotypes (Figure 2). A similar pattern was observed in shoots: genotype X24 maintained higher accumulation of Mg ($Z = 1.49$), Ca ($Z = 1.36$), and Mn ($Z = 1.89$); genotype 58 had the highest K ($Z = 0.95$); genotype 65 showed the highest Zn ($Z = 1.74$); and genotype 20 displayed slightly higher Fe levels ($Z = 0.65$). These results indicate distinct elemental accumulation patterns among genotypes, with 65 showing strong Zn affinity and X24 exhibiting strength in multiple elements (Mg, Ca, Mn), demonstrating consistent root-shoot accumulation patterns.

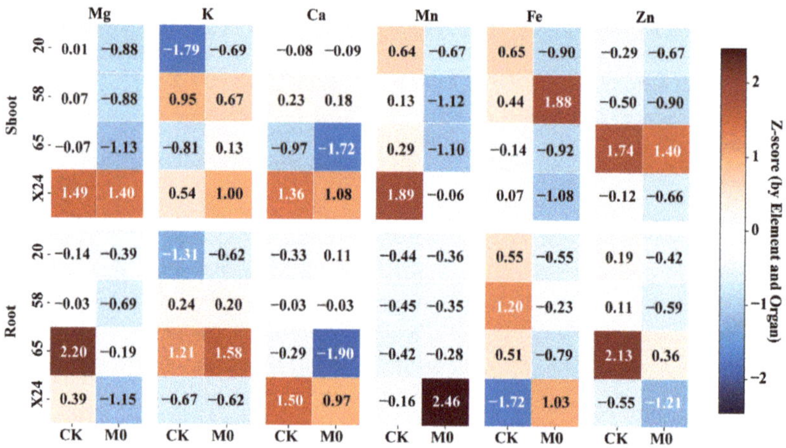

Figure 2. Heatmap of Z-scores for essential element accumulation in roots and shoots of four rice genotypes under Cd and control conditions. CK: Untreated control group (no additions); M0: 0.3 mg·L^{-1} Cd (sole Cd stress treatment).

Under Cd stress, both common and genotype-specific responses were observed (Figure 2, Table A2). Common responses included: significant reduction in root Mg and Zn ($\Delta Z < 0$) and shoot Mg, Ca, Mn, and Zn ($\Delta Z < 0$) across all genotypes; and increased root Mn accumulation ($\Delta Z > 0$), although this effect was most pronounced in genotype X24 ($\Delta Z = +2.62$). Genotype-specific responses revealed the following: Cd significantly enhanced root Fe accumulation only in X24 ($\Delta Z = +2.75$) while reducing it in others; specifically promoted shoot Fe accumulation in genotype 58 ($\Delta Z = +1.44$); and differentially affected K distribution, increasing root K in genotypes 20 and 65 but decreasing shoot K in genotype 58.

These findings demonstrate that Cd stress generally disrupts nutrient homeostasis but elicits genotype-specific adaptations, particularly in Fe and Mn accumulation, suggesting distinct genetic strategies for coping with cadmium toxicity.

3.3. *Effects of Malic Acid on Cd Accumulation in Rice Genotypes*

When rice seedlings of genotypes X24, 58, and 65 were exposed to 0.3 mg·L^{-1} Cd stress, with the addition of 0.5 or 1.5 mmol·L^{-1} malic acid, Cd accumulation in both shoots and roots significantly decreased, with varying degrees of reduction among genotypes (Figure 3).

Without malic acid, genotype X24 showed significantly lower Cd accumulation in shoots and roots compared to genotypes 58 and 65. Compared to the control (M0), 0.5 and 1.5 mmol·L^{-1} malic acid reduced shoot Cd content in genotype X24 by 36.8% and 63.2%, respectively, and root Cd content by 29.9% and 62.5%. For genotype 58, shoot Cd content decreased by 51.2% and 67.5%, and root Cd content by 25.1% and 65.3%. In genotype 65, shoot Cd content declined by 41.6% and 68.0%, and root Cd content by 29.2% and 59.5%. These results demonstrate that 0.5, 1.5 mmol·L^{-1} malic acid effectively inhibits Cd accumulation, with significant genotypic differences in sensitivity to malic acid.

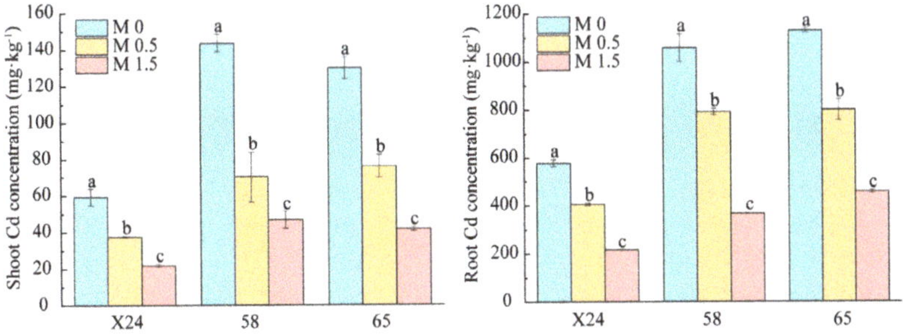

Figure 3. Effect of malic acid on cadmium accumulation and partitioning in rice root and shoot under Cd stress. M0: 0.3 mg·L^{-1} Cd (sole Cd stress treatment); M0.5: 0.3 mg·L^{-1} Cd + 0.5 mM malic acid; M1.5: 0.3 mg·L^{-1} Cd + 1.5 mM malic acid. Different letters indicate a significant difference between different genotypes at 5% level.

3.4. Effects of Malic Acid on Essential Elements in Rice Genotypes

The application of exogenous malic acid (0.5 and 1.5 mmol·L^{-1}) induced pronounced, genotype-dependent shifts in the accumulation of essential elements in both roots and shoots (two key nutritional organs) of rice seedlings (Genotypes X24, 58, and 65), as detailed by Z-score analysis (Figure 4).

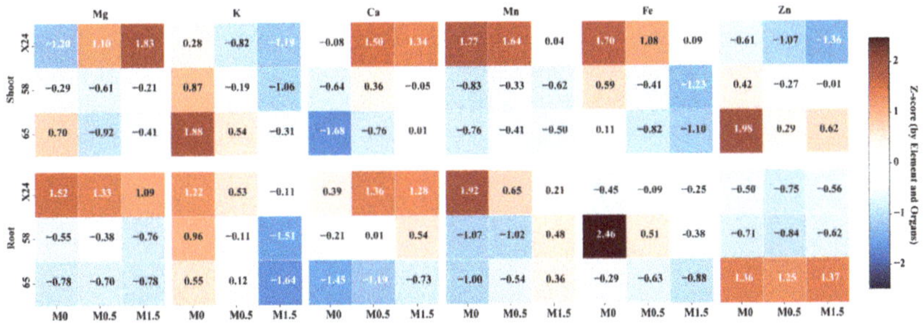

Figure 4. Heatmap of Z-scores for essential element accumulation in roots and shoots of rice genotypes X24, 58, and 65 under malic acid treatments. M0: 0.3 mg·L^{-1} Cd (sole Cd stress treatment); M0.5: 0.3 mg·L^{-1} Cd + 0.5 mM malic acid; M1.5: 0.3 mg·L^{-1} Cd + 1.5 mM malic acid.

Root responses revealed both common and genotype-specific patterns across all three genotypes (Figure 4, Table A3). A consistent regulatory effect was observed for potassium (K) and calcium (Ca): malic acid significantly suppressed K accumulation and notably enhanced Ca accumulation in roots of all genotypes. Conversely, the effects on zinc (Zn), iron (Fe), magnesium (Mg), and manganese (Mn) were highly genotype-specific. In genotype X24, malic acid application maintained relatively high Mg accumulation (Z-scores: 1.52 → 1.33 → 1.09) and slightly reduced Mn accumulation (1.92 → 0.66 → 0.21), though Mn levels remained above the overall average (Z-score > 0). In genotype 58, a notable promotion of Mn accumulation was observed specifically under the 1.5 mmol·L^{-1} malic acid treatment (Z-score: −1.07 → 0.48, shifting from below to above average), while Fe accumulation was strongly suppressed (2.46 → −0.38). Genotype 65 exhibited a similar trend of Mn promotion (−1.00 → 0.36) and Fe suppression (−0.29 → −0.88), while maintaining stably high Zn levels (Z-scores > 1.24) across all malic acid treatments.

Shoot responses demonstrated even greater genotypic divergence compared to roots (Figure 4, Table A4). Most strikingly, in genotype X24, malic acid triggered a dramatic shift in Mg and Ca homeostasis: Mg accumulation increased from below the overall average to a highly abundant state (Z-score: −1.20 → 1.83), and Ca accumulation also significantly elevated (from −0.08 to 1.34). This shift was accompanied by a concurrent reduction in the accumulation of K, Fe, and Zn. The responses of genotypes 58 and 65 were markedly different from that of X24. Both genotypes showed increased Ca accumulation (e.g., genotype 65: Z-score −1.68 → 0.01) and decreased K accumulation in shoots; however, unlike X24, their Mg, Fe, and Zn status remained largely unchanged. Notably, Mn accumulation in shoots was consistently high in genotype X24 (Z-score > 0) but persistently low in genotypes 58 and 65 (Z-score < 0) across all treatments, highlighting a fundamental genotypic difference in Mn translocation and accumulation.

In summary, exogenous malic acid consistently modulates the distribution of K and Ca in both roots and shoots across the three rice genotypes, while eliciting highly divergent, genotype-specific effects on the homeostasis of Mg, Mn, Fe, and Zn. These findings underscore that genetic background is a primary determinant of the nutritional responses of rice seedlings to organic acid amendments.

3.5. Effect of Malic Acid on Root Morphology of Rice Genotypes Under Cd Stress

This study selected two rice genotypes (58 and 65) for investigation, which were treated with 0.3 mg·L^{-1} Cd for 10 days. The rice root system was first affected by Cd toxicity, and the root length, root surface area, root volume, and root tip number of seedlings could reflect the tolerance of rice root organs to Cd.

Compared with the CK treatment, 0.3 mg·L^{-1} Cd significantly inhibited root development in rice genotype 58, reducing root length by 45.8%, root surface area by 19.8%, root volume by 6.2%, and root tip number by 26.4%. Notably, the reductions in root length and surface area reached statistical significance ($p < 0.05$, Figure 5). The addition of 0.5 and 1.5 mmol·L^{-1} malic acid effectively mitigated these Cd-induced inhibitory effects. Compared with the M0 treatment, 1.5 mmol·L^{-1} malic acid treatment resulted in significant increases of 42.8% in root length and 57.8% in root tip number (Figure 5).

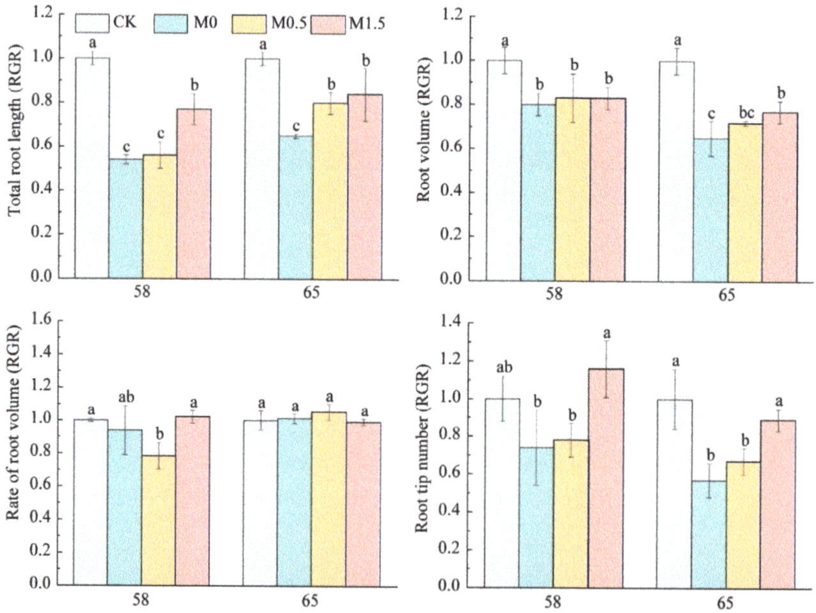

Figure 5. Malic acid rescues cadmium-impaired root growth across rice genotypes and doses. RGR (Relative Growth Rate) was calculated as the ratio of [trait value in treatment] to [trait value in control]. Values > 1 indicate growth promotion, while values < 1 indicate growth inhibition. CK:

Untreated control group (no additions); M0: 0.3 mg·L^{-1} Cd (sole Cd stress treatment); M0.5: 0.3 mg·L^{-1} Cd + 0.5 mM malic acid; M1.5: 0.3 mg·L^{-1} Cd + 1.5 mM malic acid. Different letters indicate a significant difference between different genotypes at 5% level.

Under 0.3 mg·L^{-1} Cd stress, the root length, root surface area, and root tip number of genotype 65 decreased significantly by 35.1%, 35.2%, and 42.6%, respectively, compared with CK treatment. Adding 0.5 and 1.5 mmol·L^{-1} malic acid could alleviate the inhibitory effect of Cd on the root length, root surface area, and root tip number of genotype 65 seedlings. Compared with the M0 treatment, 1.5 mmol·L^{-1} malic acid could significantly increase the root length, root surface area, and root tip number by 28.9%, 18.0%, and 55.4%. In summary, Cd inhibits root growth in both genotype 58 and 65, manifesting as a decrease in root length and root surface area parameter values and a decrease in root tip number.

Adding malic acid can significantly reduce the inhibitory effect of Cd on root development. It is speculated that adding 0.5~1.5 mmol·L^{-1} malic acid to the root system can promote the growth and development of rice roots, reduce the toxic effect of Cd on rice roots, and have a positive effect on the entire growth period of rice.

4. Discussion

Cadmium, as a toxic heavy metal, significantly inhibits plant growth and development. In our study, exposure to 0.3 mg·L^{-1} Cd (the concentration used in this experiment) significantly inhibited the growth of rice seedlings, as evidenced by reduced seedling fresh weight (Figure 6). This aligns with previous findings that Cd toxicity retards rice growth at low concentrations and severely impairs all growth stages or even causes mortality at high concentrations [45,46]. Cd influences malic acid secretion in rice roots, with low concentrations promoting and high concentrations inhibiting its production [51]. While some studies suggest that root-secreted organic acids can activate Cd in the rhizosphere and enhance its bioavailability [52–56], others report that exogenous malic acid does not promote Cd uptake [57,58], indicating complex interactions between organic acids and Cd accumulation in rice [59].

Figure 6. Effects of Cd on the growth of rice seedlings. CK: Untreated control group (no additions); 0.3 ppm Cd: 0.3 mg·L^{-1} Cd (sole Cd stress treatment).

Our comparative analysis revealed distinct genotypic patterns in Cd accumulation: genotype X24 consistently showed low Cd accumulation, whereas genotypes 20, 58, and 65 were high accumulators (Figure 1). Notably, the addition of 0.5 or 1.5 mmol·L^{-1} malic acid significantly reduced Cd accumulation in roots and shoots of genotypes X24, 58, and

65 under Cd stress, with optimal reduction at 1.5 $mmol \cdot L^{-1}$ (Figure 3). This protective effect may be attributed to malic acid's role in critical metabolic processes that inhibit Cd bioavailability and absorption [60].

A central finding of this study is that genetic background fundamentally determines how rice genotypes respond to both Cd stress and malic acid amendment, particularly regarding essential element homeostasis. While Cd is known to interfere with mineral nutrient uptake [61]—suppressing Mn and Zn in wheat roots [62] while promoting P, K, and Mn absorption [63]—our standardized Z-score analysis revealed not only genotypic variation in inhibitory effects but also a clear distinction between common stress responses and genotype-specific adaptations. Under Cd-free conditions, genotype 65 exhibited superior Zn accumulation capacity ($Z = 2.13$ in roots, $Z = 1.74$ in shoots), while genotype X24 showed prominent accumulation of multiple elements (Mg, Ca, and Mn) with consistent root–shoot distribution patterns. Under Cd stress, all genotypes showed reduced root Mg/Zn and shoot Mg/Ca/Mn/Zn ($\Delta Z < 0$), yet only genotype X24 displayed increased root Fe ($\Delta Z = +2.75$) and substantial root Mn enhancement ($\Delta Z = +2.62$), suggesting Mn absorption may be a key adaptive trait for X24 in coping with Cd toxicity.

Most significantly, our results demonstrate that malic acid's effect on nutrient homeostasis is profoundly genotype-dependent. Under 0.3 $mg \cdot L^{-1}$ Cd stress, malic acid consistently suppressed K accumulation and enhanced Ca accumulation across all genotypes—a common regulatory pattern potentially linked to conserved physiological roles in stress response. However, for other elements, responses varied dramatically by genotype: malic acid maintained high Mg and moderately reduced Mn (though still above average) in X24 roots; promoted Mn accumulation while strongly suppressing Fe in genotype 58; and enhanced Mn while maintaining high Zn levels in genotype 65. This genotypic divergence underscores that genetic background is a primary determinant of nutritional responses to organic acid amendments.

The mechanistic basis for these responses may involve multiple pathways. Previous studies have suggested that organic acids such as malic acid can form complexes with Cd, reducing its bioavailability, or compete for metal ion transporters responsible for Cd^{2+} uptake. The consistent reduction in K accumulation alongside Fe and Zn suppression suggests that malic acid may modulate non-selective cation channels [64] through which these elements and Cd enter rice cells, thereby reducing Cd transport while independently enhancing Ca uptake via specific transporters. Cd's physicochemical similarity to essential elements like K, Fe, and Zn enables it to compete for binding sites on carrier proteins [65], further complicating these interactions.

We observed striking genotypic divergence in shoot responses, particularly regarding Mg and Mn homeostasis. Malic acid triggered a dramatic shift in Mg and Ca accumulation specifically in genotype X24 shoots (Mg: $Z = -1.20 \rightarrow 1.83$; Ca: $Z = -0.08 \rightarrow 1.34$), while genotypes 58 and 65 showed minimal changes in these elements. Furthermore, Mn accumulation remained consistently high in X24 shoots but persistently low in 58 and 65 across all treatments, highlighting fundamental genotypic differences in Mn translocation efficiency.

The uniform increase in Ca accumulation across genotypes has important physiological implications for Cd stress mitigation. As calcium pectate in the cell wall matrix, enhanced Ca likely contributes to improved cell wall formation and intracellular osmotic regulation, maintaining normal physiological conditions under stress. Additionally, Ca's involvement in protein synthesis and enzymatic metabolism further supports plant growth and development under Cd stress. This mechanism may explain why malic acid significantly alleviated Cd-induced root growth inhibition in genotypes 58 and 65 (Figure 3), possibly through its participation in diverse metabolic pathways and enzyme formation that stimulate meristematic cell proliferation.

In conclusion, our findings demonstrate that 1.5 mmol·L^{-1} malic acid not only promotes root development and mitigates Cd toxicity but also modulates essential element homeostasis in a fundamentally genotype-dependent manner. The genetic background of the host plant emerges as a primary factor determining whether malic acid amendment will suppress, promote, or have minimal effect on specific essential elements, providing crucial insights for developing genotype-specific strategies for heavy metal stress management in rice. While our physiological and elemental data strongly support a role for malic acid in modulating Cd uptake and nutrient homeostasis, future studies incorporating gene expression analyses and direct measurement of metal–organic acid complexes will be essential to fully elucidate the underlying molecular mechanisms.

5. Conclusions

Malic acid serves as an effective amendment for mitigating cadmium toxicity in rice, but its efficacy and mechanistic actions are fundamentally governed by the plant's genetic background. We identified distinct Cd accumulation patterns among genotypes, with X24 exhibiting low accumulation and 58/65 being high accumulators. The key finding is that 1.5 mmol·L^{-1} MA optimally reduced Cd accumulation across these genotypes. More importantly, this study unveils a novel, genotype-specific reprogramming of essential element homeostasis induced by MA, characterized by consistent suppression of K and promotion of Ca, alongside highly divergent responses in Mg, Fe, Mn, and Zn accumulation. This genotypic variation extends to physiological responses, as MA significantly restored Cd-impaired root growth in sensitive genotypes. The dual role of MA in reducing Cd uptake and modulating nutrient equilibrium represents a significant advancement over previous studies focused solely on Cd mitigation. Therefore, the application of MA represents a promising strategy for reducing Cd bioavailability and accumulation in rice. However, given the strong genotype dependence, we emphasize the necessity of considering genetic profiles in developing MA-based agronomic practices or foliar conditioners, which constitutes a novel contribution to the field.

Author Contributions: S.Z.: Conceptualization, Methodology, Writing—original draft. Z.L.: Conceptualization, Methodology. T.L. , Y.J. and C.Q.: Data curation. Y.H. and Z.L.: Data curation. Y.Y. and W.D.: Formal analysis. G.L. and Y.Z.: Software, Visualization. Y.J. and C.Q.: Funding acquisition and Project administration. All authors have read and agreed to the published version of the manuscript.

Funding: This research was funded by the Key Technology R&D and Application for Improving Comprehensive Grain Production Capacity of Modern Agricultural Provincial Laboratory (ZY04JD05), the Heilongjiang Provincial Collaborative Innovation and Promotion System for Green and Organic Agriculture, Basic Research Support Program of Outstanding Young Teachers in Heilongjiang Provincial Undergraduate Universities (YQJH2023181), National Natural Science Foundation of China (42407423), the Heilongjiang Province Agricultural Science and Technology Innovation Span Project Agricultural Science and Technology Foundation Innovation Project (CX25JC75).

Institutional Review Board Statement: Not applicable.

Informed Consent Statement: Not applicable.

Data Availability Statement: All data generated or analyzed during this study are included in this published article.

Conflicts of Interest: The authors declare that they have no known competing financial interests or personal relationships that could have appeared to influence the work reported in this paper.

Abbreviations

The following abbreviations are used in this manuscript:

Cd	Cadmium
TCA	Tricarboxylic acid
CK	Control group
MA	Malic acid

Appendix A

Table A1. The accumulation of essential elements in rice roots and shoots of different genotypes (Mean ± SD, mg·kg^{-1}). Different letters indicate a significant difference between different genotypes at 5% level.

	Genotype	Mg	K	Ca	Mn	Fe	Zn
Root	X24	1122 ± 280 bc	12,358 ± 2461 a	3480 ± 220 a	128 ± 48 a	2316 ± 146 b	205 ± 47 b
	20	987 ± 76 c	9681 ± 1324 d	2555 ± 198 b	45 ± 11 b	4998 ± 126 a	291 ± 32 b
	58	1014 ± 79 bc	16,099 ± 778 b	2707 ± 107 b	43 ± 3 b	5757 ± 315 a	281 ± 13 b
	65	1586 ± 46 a	20,130 ± 592 a	2579 ± 148 b	50 ± 8 b	4951 ± 289 a	514 ± 47 a
Shoot	X24	6061 ± 414 a	52,747 ± 1729 a	6463 ± 479 a	676 ± 33 a	369 ± 138 b	131 ± 16 b
	20	4842 ± 60 b	45,501.7 ± 984 b	5382 ± 166 bc	469 ± 23 b	454 ± 65 a	119 ± 8 b
	58	4893 ± 439 b	54,039 ± 6137 a	5616 ± 340 b	384 ± 25 c	424 ± 101 a	104 ± 14 b
	65	4775 ± 180 b	48,533 ± 2423 b	4717 ± 141 c	411 ± 16 c	338 ± 54 b	263 ± 6 a

Table A2. The content of essential elements in rice roots and shoots of different genotypes under 0.3 mg·L^{-1} Cd stress (Mean ± SD, mg·kg^{-1}). Different letters indicate a significant difference between different genotypes at 5% level.

	Genotype	Mg	K	Ca	Mn	Fe	Zn
Root	X24	728 ± 22 d	12,568 ± 1698 c	3214 ± 157 a	907 ± 106 a	5562 ± 2089 a	129 ± 13 d
	20	922 ± 20 b	12,543 ± 257 c	2779 ± 43 b	70 ± 2 b	3700 ± 73 ab	220 ± 11 b
	58	846 ± 26 c	15,955 ± 776 b	2709 ± 61 b	71 ± 6 b	4073 ± 257 ab	201 ± 10 c
	65	975 ± 41 a	21,667 ± 1095 a	1766 ± 190 c	92 ± 41 b	3422 ± 155 b	310 ± 7 a
Shoot	X24	5987 ± 112 a	54,199 ± 885 a	6255 ± 52 a	353 ± 8 a	201 ± 44 b	93 ± 15 b
	20	4107 ± 330 b	48,909 ± 3883 a	5375 ± 450 b	252 ± 22 b	227 ± 24 b	92 ± 9 b
	58	4114 ± 301 b	53,145 ± 3525 a	5574 ± 272 b	178 ± 11 c	634 ± 335 a	76 ± 4 b
	65	3905 ± 140 b	51,482 ± 490 a	4155 ± 111 c	182 ± 15 c	225 ± 121 b	239 ± 6 a

Table A3. Effect of malic acid on the accumulation of essential elements in rice roots of different genotypes (Mean ± SD, mg·kg^{-1}). Different letters indicate a significant difference between different genotypes at 5% level.

	Genotype	Mg	K	Ca	Mn	Fe	Zn
X24	M0	728 ± 22 c	12,568 ± 1698 a	3214 ± 157 b	907 ± 106 a	5562 ± 2089 a	129 ± 13 a
	M0.5	1027 ± 52 b	6317 ± 2087 b	4634 ± 83 a	863 ± 312 a	4721 ± 369 a	97 ± 12 b
	M1.5	1121 ± 59 a	4264 ± 555 b	4487 ± 71 a	351 ± 12 b	3403 ± 1138 a	77 ± 11 b
58	M0	846 ± 26 a	15,955 ± 776 a	2709 ± 61 b	71 ± 6 c	4073 ± 257 a	201 ± 10 a
	M0.5	805 ± 59 a	9915 ± 699 b	3602 ± 672 a	230 ± 13 a	2727 ± 226 b	153 ± 11 b
	M1.5	856 ± 26 a	4974 ± 250 c	3235 ± 144 ab	139 ± 7 b	1628 ± 58 c	171 ± 25 ab
65	M0	975 ± 41 a	21,667 ± 1095 a	1766 ± 190 c	92 ± 4 c	3422 ± 155 a	310 ± 7 a
	M0.5	765 ± 61 b	14,047 ± 715 b	2596 ± 19 b	205 ± 7 a	2182 ± 98 b	192 ± 5 c
	M1.5	831 ± 52 b	9259 ± 919 c	3292 ± 167 a	176 ± 10 b	1806 ± 34 c	215 ± 4 b

Table A4. Effect of malic acid on the accumulation of essential elements in rice shoots of different genotypes (Mean ± SD, mg·kg^{-1}). Different letters indicate a significant difference between different genotypes at 5% level.

	Genotype	Mg	K	Ca	Mn	Fe	Zn
X24	M0	5987 ± 112 a	54,199 ± 885 a	6255 ± 52 a	353 ± 8a	201 ± 44a	93 ± 15 a
	M0.5	5816 ± 409 a	51,404 ± 3270 ab	7363 ± 841 a	279 ± 27 b	255 ± 106 a	73 ± 8 a
	M1.5	5602 ± 393 a	48,853 ± 2872 b	7276 ± 609 a	253 ± 7 b	231 ± 126 a	88 ± 6 a
58	M0	4113 ± 301 a	53,145 ± 3525 a	5574 ± 272 b	178 ± 11 b	634 ± 335 a	76 ± 4 a
	M0.5	4267 ± 32 a	48,831 ± 572 b	5820 ± 54 b	181 ± 3 b	343 ± 6 ab	66 ± 1 b
	M1.5	3923 ± 21 a	43,179 ± 968 c	6427 ± 171 a	269 ± 3 a	211 ± 20 b	83 ± 5 a
65	M0	3905 ± 140 a	51,482 ± 490 a	4155 ± 111 b	182 ± 15 b	225 ± 121 a	239 ± 6 a
	M0.5	3980 ± 318 a	49,770 ± 4399 a	4453 ± 381 b	209 ± 16 b	174 ± 56 a	230 ± 16 a
	M1.5	3903 ± 121 a	42,647 ± 1177 b	4977 ± 169 a	262 ± 18 a	138 ± 33 a	240 ± 19 a

References

1. Ahmad, S.; Sehrish, A.K.; Alomrani, S.O.; Zhang, L.; Waseem, M.; Noureen, S.; Ullah, I.; Tabassam, R.; Abbas, G.; Ali, S. Combined application of biochar and metal-tolerant bacteria alleviates cadmium toxicity by modulating the antioxidant defense mechanism and physicochemical attributes in rice (*Oryza sativa* L.) grown in cadmium-contaminated soil. *Plant Stress* **2024**, *11*, 100348. [CrossRef]
2. Qin, Y.; Li, Z.; Sun, J.; Xu, M.; Gu, M.; Wei, Y.; Lei, J. Manganese (II) sulfate affects the formation of iron-manganese oxides in soil and the uptake of cadmium and arsenic by rice. *Ecotoxicol. Environ. Saf.* **2023**, *263*, 115360. [CrossRef] [PubMed]
3. Yuan, S.; Chen, P.; Zhou, W.; Liu, H.; Cheng, K.; Xiao, X.; Tang, H.; Yi, Z. Response characteristics of soil Cd availability to microbes in paddy soil with long-term fertilization and its impact on Cd uptake in rice. *Sci. Total Environ.* **2024**, *957*, 177680. [CrossRef] [PubMed]
4. Lwin, C.S.; Jung, H.; Kim, M.S.; Lee, E.J.; Lee, T.G. Geochemical speciation, uptake, and transportation mechanisms of arsenic, cadmium, and lead in soil–rice systems: Additional aspects and challenges. *Antioxidants* **2025**, *14*, 607. [CrossRef]
5. Teng, Y.; Hu, J.; Sun, H.; Xiao, Y.; Guo, J.; Yu, H. Research on the impact of elevated atmospheric CO_2 concentration on Cd absorption in the cell walls of rice roots. *Process Saf. Environ. Prot.* **2025**, *200*, 107379. [CrossRef]
6. Haider, F.U.; Liqun, C.; Coulter, J.A.; Cheema, S.A.; Wu, J.; Zhang, R.; Wenjun, M.; Farooq, M. Cadmium toxicity in plants: Impacts and remediation strategies. *Ecotoxicol. Environ. Saf.* **2021**, *211*, 111887. [CrossRef]
7. Zhang, Z.; Qin, Y.; Ya, F.; Ma, Y.; Pan, G.; Peng, S.; Liang, S.; Feng, Y.; Jiang, Z.; Gu, M.; et al. Foliar spraying of a novel biosynthesized nano-selenium: Enhancing cadmium reduction in rice grains and regulating in vitro bioavailability with mechanistic insights. *J. Food Compos. Anal.* **2025**, *148*, 108280. [CrossRef]
8. Guo, R.; Ren, R.; Wang, L.; Zhi, Q.; Yu, T.; Hou, Q.; Yang, Z. Using machine learning to predict selenium and cadmium contents in rice grains from black shale-distributed farmland area. *Sci. Total Environ.* **2024**, *912*, 168802. [CrossRef]
9. Hafez, E.M.; Gao, Y.; La, H.G.; Alharbi, K.; Hamada, M.M.; Omara, A.E.-D.; Alshaal, T. Enhancing rice productivity in wastewater-irrigated saline Cd-contaminated soils using microbial-nanoparticle synergy. *Environ. Technol. Innov.* **2025**, *39*, 104253. [CrossRef]
10. Iqbal, A.; Ligeng, J.; Mo, Z.; Adnan, M.; Lal, R.; Zaman, M.; Usman, S.; Hua, T.; Imran, M.; Pan, S.; et al. Substation of vermicompost mitigates Cd toxicity, improves rice yields and restores bacterial community in a Cd-contaminated soil in Southern China. *J. Hazard. Mater.* **2024**, *465*, 133118. [CrossRef]
11. Roy, R.; Hossain, A.; Sarker, T.; Asaduzzaman, M. Alleviation of heavy metal accumulation in rice: A synergistic planetary health approach of organic amendment and phytoremediation in cadmium contaminated soil. *J. Hazard. Mater. Adv.* **2025**, *19*, 100864. [CrossRef]
12. Dou, X.; Zhang, C.; Zhang, J.; Ma, D.; Chen, L.; Zhou, G.; Duan, Y.; Tao, L.; Chen, J. Relationship between calcium forms and organic carbon content in aggregates of calcareous soils in northern China. *Soil Tillage Res.* **2024**, *244*, 106210. [CrossRef]
13. Wu, X.; Lin, Q.; Li, G.; Guo, C.; Li, L.; Wang, J. Evaluating water management efficiency in regulating cadmium and arsenic accumulation in rice in typical japonica paddy soils at varied pH levels. *Agriculture* **2024**, *14*, 407. [CrossRef]
14. Veeken, A.; Hamelers, B. Sources of Cd, Cu, Pb and Zn in biowaste. *Sci. Total Environ.* **2002**, *300*, 87–98. [CrossRef] [PubMed]
15. Li, Y.; Liu, M.; Wang, H.; Li, C.; Zhang, Y.; Dong, Z.; Fu, C.; Ye, Y.; Wang, F.; Chen, X.; et al. Effects of different phosphorus fertilizers on cadmium absorption and accumulation in rice under low-phosphorus and rich-cadmium soil. *Environ. Sci. Pollut. Res.* **2024**, *31*, 11898–11911. [CrossRef]
16. Mukta, T.A.; Hoque, M.A.; Shimo, F.J.; Islam, S. Cadmium contamination in rice and associated human health risk. *Agric. (Pol'nohospodárstvo)* **2024**, *70*, 38–52. [CrossRef]

17. El-Mamoon, M.F.; Abdel-Salam, A.A.; Abdel-Hmied, A.H.; Moursy, A.A.; Hekal, M.A. Concentrations of Cd, Co, and Pb in soil and taro plant (*Colocasia esculenta*) at various distances from a phosphate fertilizer factory. *Benha J. Appl. Sci.* **2023**, *8*, 75–78. [CrossRef]
18. Shao, X.; Yao, H.; Cui, S.; Peng, Y.; Gao, X.; Yuan, C.; Chen, X.; Hu, Y.; Mao, X. Activated low-grade phosphate rocks for simultaneously reducing the phosphorus loss and cadmium uptake by rice in paddy soil. *Sci. Total Environ.* **2021**, *780*, 146550. [CrossRef]
19. Xia, F.; Zhao, Z.; Niu, X.; Wang, Z. Integrated pollution analysis, pollution area identification and source apportionment of heavy metal contamination in agricultural soil. *J. Hazard. Mater.* **2024**, *465*, 133215. [CrossRef]
20. Wang, S.; Huang, D.; Zhu, Q.; Zhu, H.; Liu, S.; Luo, Z.; Cao, X.; Wang, J.; Rao, Z.; Shen, X. Speciation and phytoavailability of cadmium in soil treated with cadmium-contaminated rice straw. *Environ. Sci. Pollut. Res.* **2015**, *22*, 2679–2686. [CrossRef]
21. Yan, T.; Gao, Z.; Wang, T.; Lu, X.; Yang, L.; Shen, L.; Zhang, Q.; Hu, J.; Ren, D.; Zhang, G.; et al. Appropriate supply of ammonium nitrogen and ammonium nitrate reduces cadmium content in rice seedlings by inhibiting cadmium uptake and transport. *Rice Sci.* **2024**, *31*, 587–602. [CrossRef]
22. Song, J.; Song, Q.; Wang, D.; Liu, Y. Mitigation strategies for excessive cadmium in rice. *Compr. Rev. Food Sci. Food Saf.* **2023**, *22*, 3847–3869. [CrossRef]
23. Madomguia, D.; Basokdou, G.B.; Appoline Isabelle, K.W.; Dongo, P.K.; Nya, E.; Togouet, S.H.Z. Accumulation of heavy metals in maga-pouss rice fields (far-north region, Cameroon) and transfer to rice grains. *Agric. Sci.* **2024**, *15*, 311–326. [CrossRef]
24. Jiang, K.; Chen, S.; Xiang, N.; Hu, J.; Cheng, Y. Deciphering the translocation of cadmium in indica rice under high-cadmium soil conditions. *Environ. Geochem. Health* **2025**, *47*, 303. [CrossRef] [PubMed]
25. Dabgar, S.; Bhavsar, S.; Modi, N. Effects of lead and cadmium on shoot length and root length of *Cascabela thevetia* (L.) lippold. *Int. Assoc. Biol. Comput. Digest* **2023**, *3*, 301–306. [CrossRef]
26. Barman, F.; Majumdar, S.; Arzoo, S.H.; Kundu, R. Genotypic variation among 20 rice cultivars/landraces in response to cadmium stress grown locally in West Bengal, India. *Plant Physiol. Biochem.* **2020**, *148*, 193–206. [CrossRef]
27. Liu, T.; Hu, W.; Weng, L.; Deng, L.; Li, J.; Yu, J.; Zhou, Z.; Liu, Y.; Chen, C.; Sheng, T.; et al. Phenotypic and genetic dissection of the contents of important metallic elements in hybrid rice grown in cadmium-contaminated paddy fields. *Heliyon* **2023**, *9*, e19919. [CrossRef] [PubMed]
28. Xu, Q.; Wang, C.; Li, S.; Li, B.; Li, Q.; Chen, G.; Chen, W.; Wang, F. Cadmium adsorption, chelation and compartmentalization limit root-to-shoot translocation of cadmium in rice (*Oryza sativa* L.). *Environ. Sci. Pollut. Res.* **2017**, *24*, 11319–11330. [CrossRef]
29. Pan, B.; Cai, Y.; Liu, B.; Cai, K.; Lv, W.; Tian, J.; Wang, W. Abatement of Cd in rice grain and toxic risks to human health by the split application of silicon at transplanting and jointing period. *J. Environ. Manag.* **2022**, *302*, 114039. [CrossRef]
30. Hakeem, K.R.; Alharby, H.F.; Pirzadah, T.B. Exogenously applied calcium regulates antioxidative system and reduces cadmium-uptake in Fagopyrum esculentum. *Plant Physiol. Biochem.* **2022**, *180*, 17–26. [CrossRef]
31. Uraguchi, S.; Mori, S.; Kuramata, M.; Kawasaki, A.; Arao, T.; Ishikawa, S. Root-to-shoot Cd translocation via the xylem is the major process determining shoot and grain cadmium accumulation in rice. *J. Exp. Bot.* **2009**, *60*, 2677–2688. [CrossRef] [PubMed]
32. He, L.; Huang, D.; Zhang, Q.; Zhu, H.; Xu, C.; Li, B.; Zhu, Q. Meta-analysis of the effects of liming on soil pH and cadmium accumulation in crops. *Ecotoxicol. Environ. Saf.* **2021**, *223*, 112621. [CrossRef] [PubMed]
33. Liu, Y.; Zhang, C.; Zhao, Y.; Sun, S.; Liu, Z. Effects of growing seasons and genotypes on the accumulation of cadmium and mineral nutrients in rice grown in cadmium contaminated soil. *Sci. Total Environ.* **2017**, *579*, 1282–1288. [CrossRef] [PubMed]
34. Li, Y.; Zhou, W.; Xiao, H.; Xin, J.; Zhao, C.; Tian, R. Photosynthetic responses of *Pontederia cordata* to cadmium stress: Anatomical structure, ultrastructure, physiology, and gene expression. *Plants* **2025**, *14*, 1344. [CrossRef]
35. Che, S.; Wang, J.; Zhou, Y.; Yue, C.; Zhou, X.; Xu, Y.; Tian, S.; Cao, Z.; Wei, X.; Li, S.; et al. The adsorption and fixation of Cd and Pb by the microbial consortium weakened the toxic effect of heavy metal-contaminated soil on rice. *Chem. Eng. J.* **2024**, *497*, 154684. [CrossRef]
36. Huang, G.; Huang, Y.; Ding, X.; Ding, M.; Wang, P.; Wang, Z.; Jiang, Y.; Zou, L.; Zhang, W.; Li, Z. Effects of high manganese-cultivated seedlings on cadmium uptake by various rice (*Oryza sativa* L.) genotypes. *Ecotoxicol. Environ. Saf.* **2023**, *264*, 115440. [CrossRef]
37. Gulzar, U.; Hussain, A.; Hamayun, M.; Iqbal, A.; Seleiman, M.F.; Alotaibi, M.; Lee, B. Gibberellins producing endophytic *Aspergillus nidulans* DSE-2 biosorbs Cd and down-regulates *OsNRAMP5* and *OsCd1* genes to improve rice growth in contaminated soil. *Plant Physiol. Biochem.* **2025**, *221*, 109650. [CrossRef]
38. Ghouri, F.; Shahid, M.J.; Liu, J.; Lai, M.; Sun, L.; Wu, J.; Liu, X.; Ali, S.; Shahid, M.Q. Polyploidy and zinc oxide nanoparticles alleviated Cd toxicity in rice by modulating oxidative stress and expression levels of sucrose and metal-transporter genes. *J. Hazard. Mater.* **2023**, *448*, 130991. [CrossRef]
39. Shahzad, M.; Peng, D.; Khan, A.; Ayyaz, A.; Askri, S.M.H.; Naz, S.; Huang, B.; Zhang, G. Sufficient manganese supply is necessary for *OsNramp5* knockout rice plants to ensure normal growth and less Cd uptake. *Ecotoxicol. Environ. Saf.* **2024**, *288*, 117386. [CrossRef]

40. Feng, X.; Pang, Z.; Liu, J.; Peng, H.; Liang, Y.; Guo, B. Chemical and microbial mechanisms underpinning calcium-regulated suppression of cadmium bioavailability in alkaline paddy soil and cadmium accumulation in rice grain. *Ecotoxicol. Environ. Saf.* **2025**, *302*, 118620. [CrossRef]
41. Wang, S.; Dong, Q.; Wang, Z. Differential effects of citric acid on cadmium uptake and accumulation between tall fescue and Kentucky bluegrass. *Ecotoxicol. Environ. Saf.* **2017**, *145*, 200–206. [CrossRef] [PubMed]
42. Hawrylak-Nowak, B.; Dresler, S.; Matraszek, R. Exogenous malic and acetic acids reduce cadmium phytotoxicity and enhance cadmium accumulation in roots of sunflower plants. *Plant Physiol. Biochem.* **2015**, *94*, 225–234. [CrossRef] [PubMed]
43. Mnasri, M.; Ghabriche, R.; Fourati, E.; Zaier, H.; Sabally, K.; Barrington, S.; Lutts, S.; Abdelly, C.; Ghnaya, T. Cd and Ni transport and accumulation in the halophyte *Sesuvium portulacastrum*: Implication of organic acids in these processes. *Front. Plant Sci.* **2015**, *6*, 156. [CrossRef] [PubMed]
44. Su, Y.; Shi, Q.; Li, Z.; Deng, H.; Zhou, Q.; Li, L.; Zhao, L.; Yuan, S.; Liu, S.; Chen, Y. *Rhodopseudomonas* palustris shapes bacterial community, reduces Cd bioavailability in Cd contaminated flooding paddy soil, and improves rice performance. *Sci. Total Environ.* **2024**, *926*, 171824. [CrossRef]
45. Kováčik, J.; Klejdus, B.; Babula, P.; Hedbavny, J. Ascorbic acid affects short-term response of *Scenedesmus quadricauda* to cadmium excess. *Algal Res.* **2017**, *24*, 354–359. [CrossRef]
46. Guo, H.; Chen, H.; Hong, C.; Jiang, D.; Zheng, B. Exogenous malic acid alleviates cadmium toxicity in *Miscanthus sacchariflorus* through enhancing photosynthetic capacity and restraining ROS accumulation. *Ecotoxicol. Environ. Saf.* **2017**, *141*, 119–128. [CrossRef]
47. Liu, D.; Gao, Z.; Li, J.; Yao, Q.; Tan, W.; Xing, W.; Lu, Z. Effects of cadmium stress on the morphology, physiology, cellular ultrastructure, and *BvHIPP24* gene expression of sugar beet (*Beta vulgaris* L.). *Int. J. Phytoremediat.* **2023**, *25*, 455–465. [CrossRef]
48. Das, S.; Sengupta, S.; Patra, P.K.; Dey, P. Limestone and yellow gypsum can reduce cadmium accumulation in groundnut (*Arachis hypogaea*): A study from a three-decade old landfill site. *Chemosphere* **2024**, *353*, 141645. [CrossRef]
49. Wu, W.; Yang, W.; Zheng, F.; Zhang, Q.; Ma, Q.; Zhao, Y.; Luo, S.; Yang, Y.; Zeng, Q.; Deng, X. Strategic attenuation of Cd accumulation in rice through stage-specific flooding: Synergistic coordination of rhizospheric Cd bioavailability, microbial communities, and iron plaque speciation. *Environ. Pollut.* **2025**, *377*, 126455. [CrossRef]
50. Zhang, S.; Han, W.; Liu, T.; Feng, C.; Jiang, Q.; Zhang, B.; Chen, Y.; Zhang, Y. Tetracycline inhibits the nitrogen fixation ability of soybean (*Glycine max* (L.) *Merr.*) nodules in black soil by altering the root and rhizosphere bacterial communities. *Sci. Total Environ.* **2024**, *908*, 168047. [CrossRef]
51. Li, Q.; Guo, J.; Zhang, X.; Yu, H.; Huang, F.; Zhang, L.; Zhang, M.; Li, T. Changes of non-protein thiols in root and organic acids in xylem sap involved in cadmium translocation of cadmium-safe rice line (*Oryza sative* L.). *Plant Soil* **2019**, *439*, 475–486. [CrossRef]
52. Pan, J.; Guan, M.; Xu, P.; Chen, M.; Cao, Z. Salicylic acid reduces cadmium (Cd) accumulation in rice (*Oryza sativa* L.) by regulating root cell wall composition via nitric oxide signaling. *Sci. Total Environ.* **2021**, *797*, 149202. [CrossRef] [PubMed]
53. Akter, S.; Samad, R. Effects of kinetin and gibberellic acid on growth and ion transport of rice under cadmium stress. *Bangladesh J. Bot.* **2024**, *53*, 209–216. [CrossRef]
54. Chen, Z.; Huang, J.; Li, S.; Shao, J.; Shen, R.; Zhu, X. Salicylic acid minimize cadmium accumulation in rice through regulating the fixation capacity of the cell wall to cadmium. *Plant Sci.* **2023**, *336*, 111839. [CrossRef]
55. Liu, Y.; Zeng, H.; Ding, S.; Hu, Z.; Tie, B.; Luo, S. A new insight into the straw decomposition associated with minerals: Promoting straw humification and Cd immobilization. *J. Environ. Sci.* **2025**, *148*, 553–566. [CrossRef]
56. Ubeynarayana, N.; Jeyakumar, P.; Bishop, P.; Pereira, R.C.; Anderson, C.W.N. Effect of soil cadmium on root organic acid secretion by forage crops. *Environ. Pollut.* **2021**, *268*, 115839. [CrossRef]
57. Zhang, H.; Xie, S.; Du, X.; Bao, Z.; Xu, F.; Awadelseid, S.F.; Yaisamut, O. Effects and mechanisms of different exogenous organic matters on selenium and cadmium uptake by rice in natural selenium-cadmium-rich soil. *Heliyon* **2024**, *10*, e37740. [CrossRef]
58. Lu, H.; Chen, L.; Liu, J.; Tang, X.; Chen, C.; Sheteiwy, M.S.; Yu, J.; Wang, X.; Han, J.; Wu, Y.; et al. Periphytic biofilm (PB) in paddy field: A natural cadmium barrier for rice (*Oryza sativa* L.) safe production through physiological detoxification and gene regulation. *J. Hazard. Mater.* **2025**, *495*, 139136. [CrossRef]
59. Wu, J.; Guo, J.; Hu, Y.; Gong, H. Distinct physiological responses of tomato and cucumber plants in silicon-mediated alleviation of cadmium stress. *Front. Plant Sci.* **2015**, *6*, 453. [CrossRef]
60. Wang, K.; Wang, F.; Yu, Y.; Yang, S.; Han, Y.; Yao, H. Microplastics and soil microbiomes. *BMC Biol.* **2025**, *23*, 273. [CrossRef]
61. Huang, L.; Li, W.; Tam, N.F.Y.; Ye, Z. Effects of root morphology and anatomy on cadmium uptake and translocation in rice (*Oryza sativa* L.). *J. Environ. Sci.* **2019**, *75*, 296–306. [CrossRef]
62. Yoshihara, T.; Goto, F.; Shoji, K.; Kohno, Y. Cross relationships of Cu, Fe, Zn, Mn, and Cd accumulations in common japonica and indica rice cultivars in Japan. *Environ. Exp. Bot.* **2010**, *68*, 180–187. [CrossRef]
63. Fontanili, L.; Lancilli, C.; Suzui, N.; Dendena, B.; Yin, Y.; Ferri, A.; Ishii, S.; Kawachi, N.; Lucchini, G.; Fujimaki, S.; et al. Kinetic analysis of zinc/cadmium reciprocal competitions suggests a possible Zn-insensitive pathway for root-to-shoot cadmium translocation in rice. *Rice* **2016**, *9*, 16. [CrossRef]

64. Zhang, X.; Xue, W.; Qi, L.; Zhang, C.; Wang, C.; Huang, Y.; Wang, Y.; Peng, L.; Liu, Z. Malic acid inhibits accumulation of cadmium, lead, nickel and chromium by down-regulation of *OsCESA* and up-regulation of *OsGLR3* in rice plant. *Environ. Pollut.* **2024**, *341*, 122934. [CrossRef] [PubMed]
65. Chen, S.; Xu, J.; Peng, L.; Cheng, Z.; Kuang, X.; Li, D.; Peng, C.; Song, H. Cadmium accumulation in rice grains is mitigated by duckweed-like hydrophyte through adsorption and increased ammonia nitrogen. *Sci. Total Environ.* **2023**, *890*, 164510. [CrossRef] [PubMed]

toxics

MDPI

Article

Toxicological Analysis of Acetamiprid Degradation by the Dominant Strain Md2 and Its Effect on the Soil Microbial Community

Jiale Zhang [1], Xin Wang [1,*], Wanlei Yue [1], Jia Bao [1], Mengqin Yao [2] and Ling Ge [1]

1 School of Environmental and Chemical Engineering, Shenyang University of Technology, Shenyang 110870, China; 18704120001@139.com (J.Z.); m15222487370@163.com (W.Y.); z1055428700@163.com (L.G.)
2 School of Chemistry and Chemical Engineering, Guizhou University, Guiyang 550025, China; mqyao@gzu.edu.cn
* Correspondence: wangxin110870@sut.edu.cn

Abstract: Microbial degradation is acknowledged as a viable and eco-friendly approach for diminishing residues of neonicotinoid insecticides. This study reports the dominant strain of Md2 that degrades acetamiprid was screened from soil and identified as *Aspergillus heterochromaticus*, and the optimal degradation conditions were determined. Research indicated that the degradation of Md2 to 100 mg/L acetamiprid was 55.30%. Toxicological analyses of acetamiprid and its metabolites subsequently revealed that acetamiprid and its metabolites inhibited the germination of cabbage seed, inhibited the growth of *Escherichia coli*, and induced the production of micronuclei in the root tip cells of faba beans. Based on the analysis of metabolic pathways, it has been determined that the primary metabolic routes of acetamiprid include N-demethylation to form IM-2-1 and oxidative cleavage of the cyanoimino group to produce IM-1-3. Using 16S rRNA high-throughput sequencing, the results showed that acetamiprid and Md2 elevated the relative abundance of *Acidithiobacillus*, *Ascomycetes*, and *Stramenobacteria*, with increases of 10~12%, 6%, and 9%, respectively, while reducing the relative abundance of *Acidobacteria*, *Chlorobacteria*, *Ascomycetes*, and *Sporobacteria*, with decreases of 15%, 8%, 32%, and 6%, respectively. The findings will facilitate the safety evaluation of the toxicological properties of neonicotinoid insecticides, their biodegradable metabolites, and associated research on their degradation capabilities.

Keywords: neonicotinoid insecticide; biodegradation; microbial community; metabolic route

1. Introduction

Acetamiprid, one of the most significant neonicotinoid pesticides, has been extensively advocated and employed in agricultural crop protection globally [1,2]. Acetamiprid targets insect nicotinic (acetylcholine) receptors (nAChR), causing abnormal excitation in pests leading to complete convulsions and paralysis, ultimately resulting in their death [3]. It has been used to control Hemiptera, mainly aphids, tassel-winged insects and Lepidoptera, to protect a wide range of crops, especially vegetables, fruits, and tea [4]. Due to its high insecticidal efficacy and lack of cross-resistance with conventional long-acting insecticides, it has been widely used over the past two decades [5]. Neonicotinoids are efficient insecticides with low toxicity, but their long-term usage has generated extensive environmental problems [6]. Some reports indicate that there is a potential toxicity risk to nontarget organisms (bees, birds, and mammals) [7] and a threat to their survival [8]. Other studies have suggested that acetamiprid may decrease human fertility [9].

Moreover, besides spraying on crop surfaces, root irrigation, and seed treatment, considerable amounts of the pesticide eventually enter the soil [10–12]. Extensive application of acetamiprid negatively impacts soil biochemical characteristics and microbial activity [13,14]. It can reduce microbial diversity, alter the community structure of soil microorganisms, affect soil bacterial abundance [15], and significantly inhibit soil respiration and phosphatase activity [16]. These impacts have prompted researchers to actively explore rapid and effective remediation strategies to mitigate these adverse environmental effects. Previous studies have shown that the microbial degradation of neonicotinoids is considered to be the most efficient and environmentally friendly in situ repair pathway [17]. Using potential and degradative microorganisms, which can grow and survive under high-stress concentrations of insecticides, offers a possible opportunity for the remediation of toxic pollutants from contaminated environments [18–22].

To this end, the present study undertook the following specific investigations: Initially, it explored the screening and identification of the dominant strain Md2 from soil that can effectively degrade acetamiprid, and how to screen out strains with efficient degradation ability from a large number of soil microorganisms has been a focus of attention for many researchers and scholars. Secondly, the dominant strain was utilized to degrade 100 mg/L of acetamiprid. The effects of pH, temperature, and inoculum size on the degradation rate were analyzed using the response surface methodology, and the optimal degradation conditions were determined. Subsequently, the toxicological properties of acetamiprid and its metabolites were assessed. The changes in toxicity during the initial and metabolic stages of acetamiprid degradation and their impact on biological systems were determined. Additionally, the metabolic pathways of acetamiprid were analyzed, which not only aids in a deeper understanding of the degradation process but also helps evaluate the potential environmental and biological impacts of its metabolites. Finally, the impact of the acetamiprid-degrading dominant strain Md2 on microbial diversity and community composition in contaminated soil was investigated using Illumina MiSeq high-throughput sequencing technology. This area has received relatively little attention so far. Simulating and assessing the effects of dominant strains on soil ecosystems under experimental conditions is of significant importance for environmental protection and agricultural sustainability. This study comprehensively tracked the process of microbial remediation of pesticide-contaminated soil, aiming to address the critical issue of pesticide residues in soil through the use of dominant strains. The findings are expected to provide scientific information to support the rational use of insecticides.

2. Materials and Methods

2.1. Chemicals and Materials

Acetamiprid (99%) was purchased from Shandong Zhongnong United Biotechnology, Inc. (Jinan, China). Methanol and acetonitrile (HPLC grade) were purchased from Thermo Fisher Scientific (China), Inc. (Beijing, China). Agarose was purchased from Beijing Aobixing Biotechnology, Inc. (Beijing, China). All other chemicals and reagents were of analytical grade and purchased from China National Medicine Chemical Reagent Inc. (Huaian, China). The composition of Luria-Bertani (LB) liquid medium was as follows: 10 g peptone, 5 g yeast extract, 5 g NaCl at pH 6, and distilled water to 1 L. Solid medium was prepared by incorporating 1.5% agarose into the previously described liquid medium. The inorganic salt medium (L^{-1}) comprised 4 g Na_2HPO_4, 1.5 g KH_2PO_4, 1 g NH_4Cl, 0.2 g $MgSO_4$•$7H_2O$, 0.02 g $CaCl_2$, 0.03 g $FeSO_4$•$7H_2O$, 1.0 g $NaNO_3$, and 1 mL of trace elements solution at pH 6. Malt extract liquid medium (L^{-1}) was prepared using 20 g malt extract. All media were sterilized at 121 °C for 30 min.

2.2. Enrichment, Isolation, and Characterization of the Acetamiprid-Degrading Strain

In this study, the soil samples used for strain screening were collected from the 5–15 cm surface layer of soil in areas of the campus that had not been previously treated with acetamiprid. All soil samples were taken at five randomly selected points, impurities were removed, and the samples were thoroughly mixed and placed in sealed polyethylene bags. The experimental soil samples were brown soil from the campus that had not been treated with acetamiprid (pH 7.9, organic matter content 1.67%, total salt content 0.11 ms/cm). The surface soil was air-dried in a cool place, gently ground, and sieved through a 40-mesh sieve. An acetamiprid biodegradation system was constructed using five grams of sieved soil and 100 mL of solution containing 20 mg/L acetamiprid at pH 7.0, in which acetamiprid served as the sole source of carbon, nitrogen, and energy. The system was incubated in a rotary shaker at 150 rpm at 30 °C for 7 days [23]. Every 7 days, 10% inoculum was transferred to a fresh medium until the acetamiprid concentration reached 200 mg/L. The 10 mL mixed bacterial solution was subsequently transferred to an inorganic salt medium containing 90 mL 100 mg/L acetamiprid and incubated in a rotary shaker for 7 days. The degradation rate of the formed mixed bacterial consortia was determined using high-performance liquid chromatography (methanol:water = 60:40 (V/V), column temperature 40 °C, flow rate 1 mL/min) [24]. The bacterial consortia exhibiting superior degradation effects were further isolated and purified. The degradation rate of each individual pure strain was then measured, and the strain with the highest degradation rate was selected to be numbered and preserved for subsequent studies.

The isolated strain was designated Md2 and analyzed through 16S rRNA gene sequencing. Cell morphology was observed under an optical microscope using cells from exponential growth. DNA was extracted using a Genomic DNA Extraction Kit [25], and the universal primers ITS1 (TCCGTAGGTGAACCTGCGG) and ITS4 (TCCTCCGCTTATTGATATGC) were used to amplify the ITS region. PCR products were recovered using an AxyPrepDNA gel Recovery Kit and DNA sequenced. The PCR amplification products were subsequently validated using agarose gel electrophoresis and compared with the NCBI 16S database [26] to obtain the strain with the highest homology to Md2.

2.3. Optimization of Acetamiprid-Degrading Conditions

The dominant strain Md2 underwent cultivation in LB liquid medium with 100 mg/L acetamiprid and was incubated on a shaker at 150 rpm for 7 days. To evaluate bacterial proliferation, samples were taken at 12 h intervals, followed by analysis of the cell-free supernatants for acetamiprid degradation [27]. The response surface method (RSM) was used to optimize the degradation conditions of acetamidine by strain Md2. Based on the methodology of Q. Zhai et al., the significant variables influencing the biodegradation of acetamiprid were optimized using the Box–Behnken design. A total of 17 experiments were conducted in a random order [24]. The concentration of acetamiprid was adjusted to 100 mg/L, with the optimal ranges of three factors selected for acetamiprid biodegradation, i.e., pH (6~8), temperature (25~35 °C), and inoculum size (0.1~0.3 g) [24,28,29]. The collected samples were analyzed using HPLC.

2.4. Toxicity Analysis of Acetamiprid and Its Metabolites

2.4.1. Effects of Acetamiprid and Its Metabolites on Cabbage Seed

Seed germination experiments serve as an excellent indicator for assessing the impact of pollutants [30]. The experiment concentrated on cabbage seeds to analyze the toxicity of acetamiprid and the toxicity of its metabolites within 14 days of degradation by the dominant strain. The Petri dishes were wrapped in newspaper and sterilized with steam at 121 °C for 30 min. Under aseptic conditions, UV-sterilized filter paper was cut and placed into the Petri dishes. Different treated inorganic salt media were evenly applied to moisten the filter paper. The top and bottom of the Petri dishes were lined with a thin layer of paper, and the corresponding solutions were sprayed to maintain humidity. The experiment involved three treatment groups: (A) sterilized (121 °C 30 min) inorganic salt medium with the addition of 5 mL acetamiprid (concentrations of 5 mg/L, 10 mg/L, 20 mg/L, 50 mg/L, and 100 mg/L); (B) sterilized inorganic salt medium supplemented with its corresponding concentration of metabolites; (C) blank control group CK (sterilized inorganic salt medium). Three parallel experiments were conducted for each group. A specific amount of cabbage seeds was evenly scattered in the Petri dishes and subsequently stored in a steady-temperature incubator at 25 °C in darkness. Samples were taken at 24 h intervals, with 5 days as the testing cycle. The seed germination rate was determined by counting the number of germinated seeds, and germination shoot length was measured.

2.4.2. Effects of Acetamiprid and Its Metabolites on *Escherichia coli*

Escherichia coli is frequently used as a classic model organism in pesticide toxicology studies due to its ease of cultivation and manipulation, as well as its rapid growth cycle [31]. Therefore, *Escherichia coli* was utilized to evaluate the toxicity of acetamiprid (before/after) biodegradation by the dominant strain in this study. A small amount of preserved *E. coli* was selected, inoculated into LB solid medium, and incubated at 37 °C and 150 rpm for 2 days. After three rounds of subculturing, the bacteria were inoculated into LB liquid medium and incubated at a constant temperature for 18 h to prepare the activated *E. coli* stock solution [32]. The experiment involved three treatment groups: to freshly prepared sterilized LB liquid medium, (A) add 10 mL inorganic salt solution of the dominant strain degrading acetamiprid (100 mg/L) for 14 d and 1 mL activated *E. coli* stock solution with OD600 value of 0.5; (B) add 10 mL sterilized inorganic salt solution containing 100 mg/L acetamiprid and 1 mL of identical *E. coli* stock solution; (C) blank control group CK, 10 mL of sterilized inorganic salt solution and 1 mL of identical *E. coli* stock solution. Throughout the experiment, the absorbance at OD600 of the *E. coli* solution was measured at intervals (2 h, 4 h, 6 h, 8 h, 10 h, and 12 h) using a UV-visible spectrophotometer to monitor *E. coli* growth. Three parallel experiments were conducted for each group.

2.4.3. Micronucleus Test

Faba bean seeds of uniform size were washed and placed in a 25 °C light-proof incubator. They were immersed in deionized water for 36 h, with the water changed every 12 h. The fully soaked seeds were then placed in a dissecting tray lined with moist absorbent cotton and gauze. The seeds were germinated in the dark at 25 °C for 12–24 h. When the embryonic roots reached a length of 1.5~2 cm, each Petri dish received four seeds, which were then exposed to acetamiprid solution in varying concentrations of 5 mg/L, 10 mg/L, 20 mg/L, 50 mg/L, and 100 mg/L for 5 h [33]. The metabolite treatment solution post-degradation of acetamiprid by the dominant strain at a corresponding concentration was used as a control experiment. Post-treatment, the seeds and root tips underwent a triple rinse with deionized water and were incubated in a constant-temperature chamber at 25 °C for 24 h, aiding in the repair and growth of root tip cells. Subsequently, the cut 1 cm long root tips were fixed in 2 mL Carnot's fixative for 36~48 h. Rinsing with SO_2 solution and deionized water resulted after the Feulgen staining process. Slides were fabricated by the pressing method, and three root tip cells removed from each treatment described above were visualized under a light microscope, with 1000 cells counted per root tip. Ultimately, the process involves identifying micronuclei, counting MCN frequencies per 100 tetrads (MCN‰), numbering mitotic cells, and calculating the mitotic index (MI). The results were obtained and expressed as mean ± standard error.

2.5. Bioremediation of Acetamiprid-Contaminated Soil by Dominant Strain Md2

To investigate the degradation ability of Md2 on acetamiprid in simulated contaminated soil, 20 mg of acetamiprid was dissolved in 50 mL of methanol to prepare a 400 mg/L acetamiprid–methanol solution. An amount of 50 mL of this solution was evenly sprayed onto 1 kg of naturally air-dried soil, mixed thoroughly, and dried in a dark fume hood for 24 h until the methanol evaporated. This process yields simulated contaminated soil with an acetamiprid concentration of 20 mg/kg. An amount of 0.1 g of Md2 was separately inoculated into sterilized (121 °C 30 min) and non-sterilized soil samples through the plate punching method, ensuring thorough mixing and dispensing into small flowerpots. The experiment included four treatment groups: (A) W-D: non-sterilized soil with only acetamiprid added; (B) M-D: sterilized soil with only acetamiprid added; (C) W-DJ: non-sterilized soil with Md2 inoculated and acetamiprid added; (D) M-DJ: sterilized soil with Md2 inoculated and acetamiprid added. Subsequently, the soil was placed in a constant-temperature incubator with the cultivation conditions: a 12 h day cycle (25 °C, light intensity of 40 lx) and a 12 h night cycle (20 °C, no light). Throughout the study, sterile water was regularly added to maintain the soil moisture at 20% of its water-holding capacity [34]. At specific intervals (0, 3, 5, 7, 10, and 14 days), 5 g of cultured soil samples were obtained to measure the degradation rate, with three parallel experiments per group.

2.6. Analysis of Microbial Communities by High-Throughput Sequencing

DNA extraction from soil samples was performed using the FastDNA Spin Kit (MP Biomedicals, Santa Ana, CA, USA). Validation of the isolated DNA was conducted through 0.8% agarose gel electrophoresis, while the quality of DNA was evaluated using a UV spectrophotometer. DNA samples were preserved at −20 °C to facilitate further PCR amplification. For PCR amplification, the fungal ITS region of the 16S rRNA gene was amplified using universal primers TCCGTAGGTGAACCTGCGG and TCCTCCGCTTATTGATATGC. The PCR cycling procedure entailed an initial denaturation at 98 °C for 30 s, followed by 27 cycles of 15 s at 98 °C, 30 s at 50 °C, and 30 s at 72 °C, and a final extension at 72 °C for 5 min [35]. The PCR products underwent purification using a 2% agarose gel DNA Purification Kit before being dispatched to PacBio in Shanghai, China, for sequencing.

Community DNA segments underwent bipartite sequencing utilizing the Illumina MiSeq system. The paired-end sequences were merged using FLASH (version 1.2.11, https://ccb.jhu.edu, accessed on 23 January 2023) and assigned to each sample based on unique barcodes and primers using Mothur software (version 1.35.1, http://www.mothur.org, accessed on 27 February 2023). Sequence analysis was performed using USEARCH software (version 8.0.1517, http://www.drive5.com/usearch/, accessed on 27 February 2023). Based on ≥97% similarity, sequences were assigned to the same operational classification unit (OTU). OTUs were categorized using the RDP classifier (http://rdp.cme.msu.edu/, accessed on 27 February 2023) with 80% confidence to obtain species-level taxonomy information for each OTU. Subsequent analysis of α-diversity and β-diversity was performed using QIIME (version 1.9.1) and R software (version 2.15.3) based on the normalized data output [23]. Community diversity was estimated using the Chao1, observed species, Shannon, and Simpson indices. Principal coordinates analysis (PCoA) was used to visualize intergroup distances and correlations in bacterial community structure.

3. Results and Discussion

3.1. Identification of Acetamiprid Degrading Strain

The dominant strains Md2 that utilized acetamiprid as the sole carbon and nitrogen source were isolated from vegetation soil. The fungus Md2 is distinguished by its radial peduncles of conidiophores that are colorless or mildly yellowish, featuring green spherical conidiophores (Figure 1). Culturing in malt extract medium, Md2 appeared fluffy and initially white, turning pale green after 1~2 days and subsequently turning to ash green within 3~5 days. Certain colonies formed droplets that were either colorless or yellowish, while the reverse side of the colony exhibited a light-yellow hue. Sequencing of the 16S rRNA gene revealed 99.63% sequence identity with *Aspergillus versicolor*. Based on its morphology and genetic sequencing data, Md2 was identified as *Aspergillus versicolor*.

3.2. Optimization of Acetamiprid Degrading Conditions

A multiple linear regression analysis was applied to the experimental data for three factors (pH (X_1), temperature (X_2), and inoculum size (X_3)). Each factor varied at three levels (−1, 0, and 1). A polynomial quadratic equation was found to represent the percentage of acetamiprid degradation, as given by the following mathematical expression:

$$Y = 56.58 - 4.09X_1 + 2.72X_2 - 0.047X_3 + 0.92X_1X_2 - 1.65X_1X_3 + 0.50X_2X_3 - 16.58(X_1)^2 - 12.87(X_2)^2 - 9.50(X_3)^2$$

where Y is the degradation rate of Md2 to acetamiprid, the regression coefficient $R^2 = 0.9824$, and $p < 0.0001$, indicating the accuracy of the employed model and good prediction of the response. The p-values for the three factors were as follows: pH, $p = 0.0031$; temperature, $p = 0.0220$; and inoculum size, $p = 0.9606$. Among these, influence size: pH > temperature > inoculum size. Through the analysis of contour plots, it was determined that the optimal degradation rate of acetamiprid is observed at intermediate values of pH, temperature, and inoculum size (Figure 2). The results indicated that the optimal conditions for Md2 to degrade acetamiprid were pH 6.8, a temperature of 30 °C, and an inoculum size of 0.2 g. According to the model fit, 55.30% of the acetamiprid degradation rate of 100 mg/L by Md2 within 7 days was obtained, which closely matches the predicted value of 56.96%, demonstrating the robustness of the response surface model in fitting the experimental data.

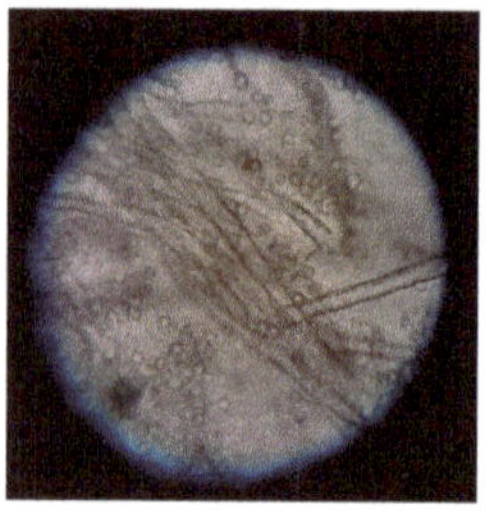

Figure 1. Optical microscope image of strain Md2 on malt powder culture medium (×400).

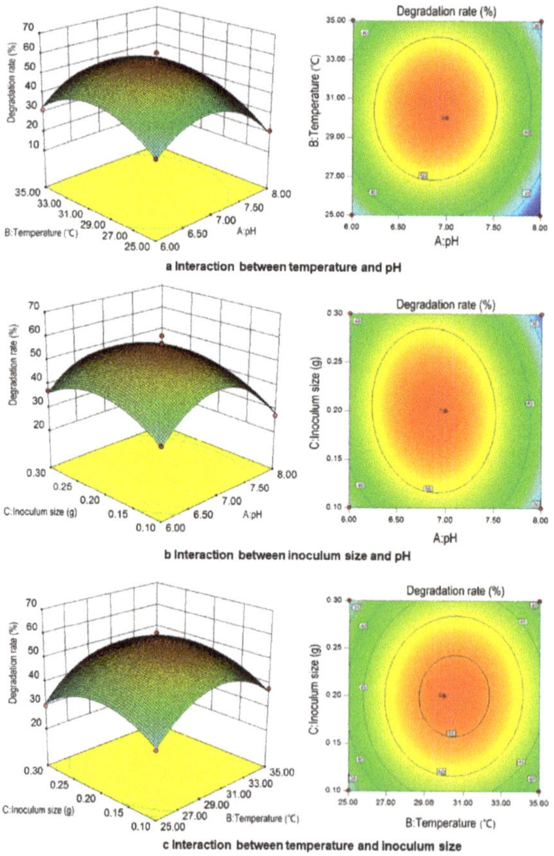

Figure 2. Response surface diagram of culture condition interaction with degradation of acetamiprid by Md2.

3.3. Biotoxicity Analysis of Acetamiprid and Its Metabolites

3.3.1. Effects of Toxicity on Cabbage Seeds

This study examines the impact of acetamiprid on the germination and seedling growth of cabbage to evaluate its toxicity to crops. Similarly, H. Qi et al. explored the impact of neonicotinoids on the growth of cabbage. This experiment found that acetamiprid and its metabolites can inhibit the seed germination and germination length of cabbage (Figure 3), with acetamiprid showing a more significant effect, particularly at a concentration of 100 mg/L, significantly inhibited seed sprouting. Previous studies have found that several pesticides, including acetamiprid, alpha-cypermethrin, and pyriproxyfen, inhibit seed germination in various plants to varying degrees [24]. Such suppression happens because specific concentrations of acetamiprid and various pesticides hinder seed respiration, affecting the germination process and resulting in inhibited seed germination, thereby causing plant seeds to suffer from phytotoxicity [30]. Compared to the metabolite group, the seed germination rate in the acetamiprid group was 18.00% to 28.66% lower, indicating that the biotoxicity of the metabolites is lower than that of acetamiprid. Examining the germination length within 5 days (Figure 3b,c) indicates that germination length decreases with increasing concentrations of acetamiprid and its metabolites, while the seedling length in the group treated with 5 mg/L of metabolites was similar to that of the control group. These results suggest that pesticides and products of their metabolism can potentially accumulate and have additive and chronic effects [36,37]. In addition, the results find that the toxicity of acetamiprid was reduced after biodegradation by the dominant strain Md2, decreasing environmental residues and providing an eco-friendly solution for soil remediation.

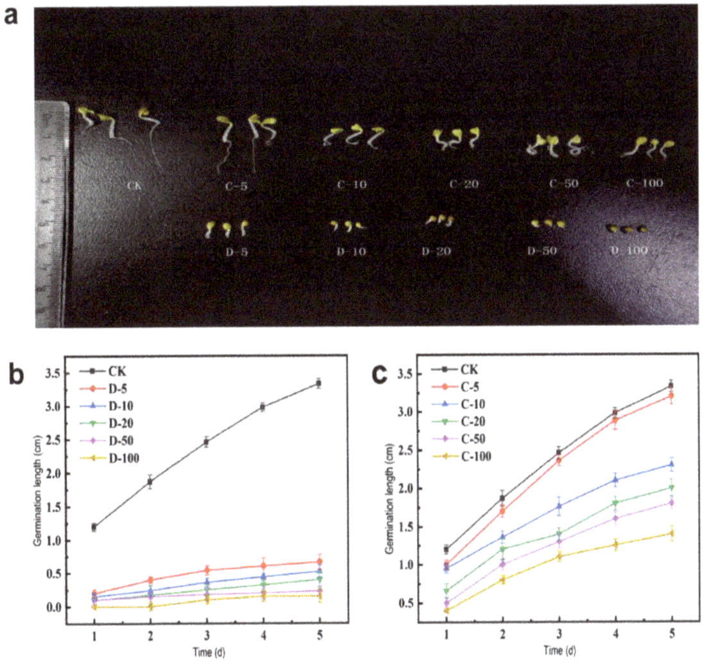

Figure 3. (**a**) Effect of acetamiprid and its metabolites on seed germination of cabbage; effects of acetamiprid (**b**) and its metabolites (**c**) on the germination length of cabbage seeds.

3.3.2. Effects of Toxicity on *E. coli*

The toxicity analysis of *Escherichia coli* (*E. coli*) was conducted over a 0–12 h period, with measurements taken every 2 h to evaluate the physiological impact of acetamiprid on bacterial growth. Figure 4 illustrates that introducing 100 mg/L acetamiprid significantly inhibits the growth rate of *E. coli*, corroborating findings by Saurabh Bhatti et al. [32]. Furthermore, the metabolite group shows some initial inhibition of *E. coli* growth within the first 6 h, followed by a diminishing inhibitory effect, with growth conditions approaching those of the blank control group after 10 h. These results indicate that acetamiprid decreases the growth rate and activity of *E. coli*, significantly suppressing bacterial growth. However, the metabolites resulting from degradation by the dominant strain Md2 exhibit a transient inhibitory effect on *E. coli* growth within a certain timeframe, with their toxicity significantly lower than that of acetamiprid. This phenomenon may be attributed to the toxic effects of acetamiprid or inhibitory intermediate metabolites on the growth metabolism of *E. coli*, thereby inhibiting its growth. This will provide a theoretical and practical foundation for future risk assessments of acetamiprid's impact on biological systems.

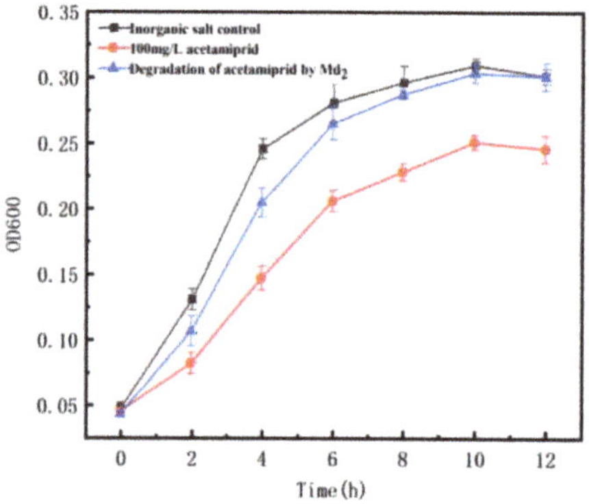

Figure 4. Effects of acetamiprid and its metabolites on *Escherichia coli*.

3.3.3. Micronucleus Test

Utilizing the micronucleus test, the degree of DNA damage in faba bean root tip cells was assessed to predict the genetic toxicity caused by acetamiprid and its metabolites, as demonstrated by MCN‰ and MI (Table 1). The study observed root tip cell micronuclei using an optical microscope (Nikon YS2 Alphaphot, Tokyo, Japan) at 400 times magnification. Results from the study indicate that the MCN‰ and MI of faba bean root tip cells increase with higher concentrations of acetamiprid and its metabolites. Similar findings by Li Wenyi et al. [33] demonstrated an increase in MCN‰ with increasing exposure time of faba beans to the same concentration of acetamiprid at 18, 24, and 48 h. These results suggest that acetamiprid may cause damage to cell chromosomes, resulting in the formation of more micronuclei during cell division. Conversely, the metabolite group exhibited significantly lower MCN‰ and MI. The experimental results show that faba bean root tip cells were hypersensitive to acetamiprid-induced micronuclei. This effect is attributed to the strong mutagenic effect of acetamiprid on the root tip cells of faba bean seeds, which interferes with the division of the spindle in mitotic cells in the plant, thus affecting cell division [38]. Furthermore, concentrations exceeding a certain threshold of acetamiprid impair the repair ability of the cell itself, causing the spindle apparatus to lag and remain in the cytoplasm as micronuclei. Recent research has established that pesticides not only act as chromosome breakers but also disrupt spindle apparatus function [39]. Therefore, acetamiprid induces micronuclei formation in faba bean root tip cells, causing severe damage to chromosome and gene structures, with greater concentrations resulting in more significant genetic damage. Importantly, the results confirm that the dominant strain Md2 effectively mitigates the toxicity of acetamiprid.

Table 1. MCN‰ and MI in faba root tip cells induced by acetamiprid and metabolites.

Concentration (mg/L)	MCN‰ of Acetamiprid Group	MI of Acetamiprid Group	MCN‰ of Metabolite Group	MI of Metabolite Group
0	0.14 ± 0.14 f	0.20 ± 0.12 f	0.20 ± 0.11 f	0.15 ± 0.15 f
5	4.50 ± 0.40 e	4.82 ± 0.58 e	0.94 ± 0.14 e	1.23 ± 0.11 e
10	5.76 ± 0.40 d	8.34 ± 0.43 d	1.64 ± 0.11 d	2.16 ± 0.12 d
20	11.09 ± 0.22 c	10.84 ± 0.51 c	3.03 ± 0.29 c	3.15 ± 0.25 c
50	13.24 ± 0.31 b	13.61 ± 0.73 b	3.85 ± 0.11 b	4.45 ± 0.22 b
100	15.99 ± 0.25 a	16.28 ± 1.05 a	4.68 ± 0.40 a	5.81 ± 0.43 a

The data in the table are the mean ± standard error. The six different letters a, b, c, d, e, and f indicate significant differences between these six groups in the statistical process.

3.4. Analysis of the Degradation Mechanism of Acetamiprid

Acetamiprid ((E)-N[1]-[(6-chloro-3-pyridyl)methyl]-N[2]-cyano-N[1]-methylacetamidine, $C_{10}H_{11}ClN_4$, AAP) is a neonicotinoid insecticide. Figure 5 shows the possible major metabolic pathways of acetamidine. In this study, the main metabolic pathway of acetamiprid is the oxidative cleavage of the pharmacophore -N=C-CN of AAP by Aspergillus versicolor Md2 to produce IM-1-3 (N-[(6-chloropyridin-3-pyridyl)methyl]-N-methylacetamide). This intermediate undergoes further oxidation steps and ultimately mineralizes into CO_2 and H_2O. At the same time, *Aspergillus versicolor* Md2 demethylates AAP to produce N-demethylated acetamiprid IM-2-1 (N[1]-[(6-chloro-3-pyridyl)methyl]-N[2]-cyanoacetamidine); the procedure is related to the cytochrome P450 [40]. After demethylation, IM-2-1 is further degraded to IC-0 (6-chloronicotinic acid), which is eventually mineralized to CO_2 and H_2O. The main metabolic pathway for the degradation of AAP by *Aspergillus versicolor* Md2 is the production of IM-1-3 and IM-2-1, which are one magnitude less biologically active than AAP [40]. IM-1-3 are almost biologically inactive owing to the lack of the pharmacophore -N=C-CN [41]. Therefore, the metabolism of AAP by *Aspergillus versicolor* Md2 is a process with decreased biotoxicity.

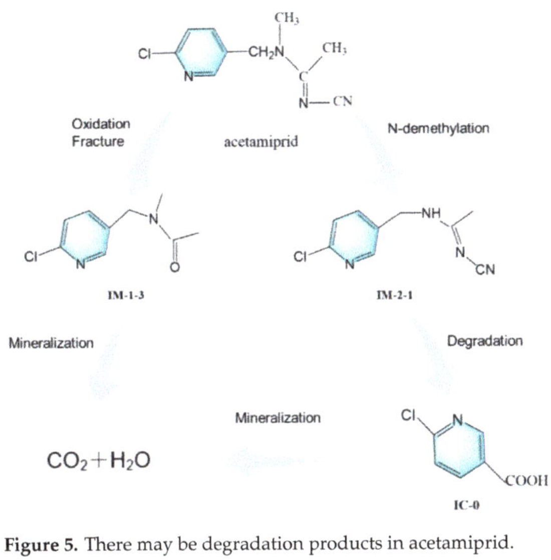

Figure 5. There may be degradation products in acetamiprid.

3.5. Effect of Md2 on the Degradation of Acetamiprid in Soil and Microbial Community Diversity

Pesticides permeate the soil through methods such as spraying, root irrigation, and seed coating. Although these chemicals are not specifically intended for the soil, the soil ecosystem quietly endures their repercussions, prompting soil microorganisms to enact a range of responses against external interference. These responses include changes in microbial diversity and community structure, the proliferation of favored populations, and even the gradual decline of particular microbial communities that are incapable of withstanding the exposure.

3.5.1. Degradation of Acetamiprid by Dominate Strain Md2

The degradation ability of Md2 towards acetamiprid was further evaluated based on the residual content of acetamiprid in sterilized and unsterilized soils treated with Md2 (a comparison of average degradation rates is shown in Figure 6). On the third day, the degradation rate of Md2 was less than 10%, possibly because excessive concentrations of pesticides may cause toxic effects [42] and decrease degradation capability. However, the degradation rate of Md2 exhibited a gradual increase with time. This suggests that the dominant strain Md2 gradually adapted to the soil environment containing acetamiprid. The unsterilized soil with added Md2 exhibited stronger degradation capability, reaching the highest degradation rate of 51.27% by the 14th day. This demonstrates that Md2 can proliferate steadily in soil, a result further corroborated by fungal community composition analysis. Furthermore, higher degradation of acetamiprid by Md2 was observed in unsterilized soil (51.27%) compared to sterilized soil (37.08%), which may be due to the presence of other types of natural acetamiprid-degrading microorganisms in the soil that metabolize acetamiprid synergistically, thus aiding in soil remediation [43].

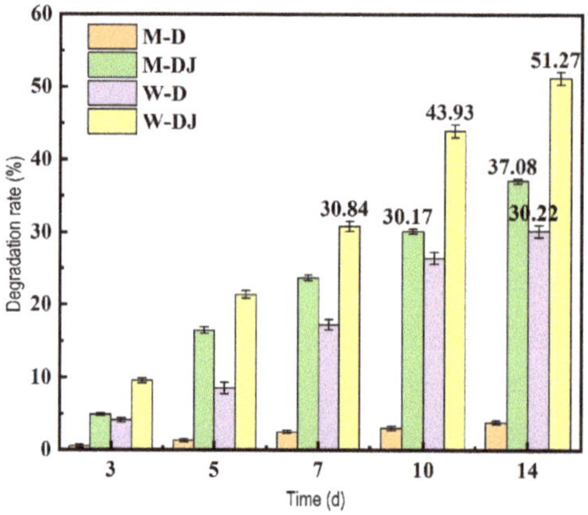

Figure 6. Degradation rate of acetamiprid in soil samples under different treatments.

3.5.2. Alpha Diversity Index Analysis

Alpha diversity was used to evaluate changes in bacterial and fungal diversity in soil communities (Table 2). Compared to the CK, the alpha diversity index of the soil decreased considerably under the influence of acetamiprid and its metabolites, demonstrating that acetamiprid resulted in a reduction in the number, variety, and diversity of bacterial and fungal species in the community. In the bacterial community (Figure 7a), the Chao1 and observed species indices notably decreased in the acetamiprid treatment group, indicating a reduction in bacterial community abundance. In the fungal community (Figure 7b), the alpha diversity indices decreased with acetamiprid treatment, indicating a reduction in fungal community abundance and diversity. However, the Shannon and Simpson indices of the group after being treated with Md2 were restored to levels similar to CK, suggesting that Md2 may partially degrade acetamiprid, leading to significant microbial diversity recovery. From the experimental results, it emerges that acetamiprid significantly diminished the alpha diversity of bacteria and fungi in soil samples. This might be related to some potentially degrading bacteria in the soil affecting acetamiprid [44]; such bacteria are capable of utilizing acetamiprid as a substrate, which could induce an increase in their population size in soils, thereby intensifying competitive selection among microorganisms [45], resulting in a decrease in alpha diversity indices.

Table 2. Index table of alpha diversity of bacteria and fungi in soil samples supplemented with acetamiprid and dominant strains.

	Group	Chao1	Observed-Species	Shannon	Simpson
Bacteria	CK	6038.21	4978.20	11.28	0.99
	D	3626.98	3126.33	8.19	0.98
	DJ	3417.11	3884.23	8.44	0.98
Fungi	CK	899.07	891.87	6.8	0.97
	D	369.96	365.83	1.7	0.28
	DJ	666.96	666.13	7.47	0.98

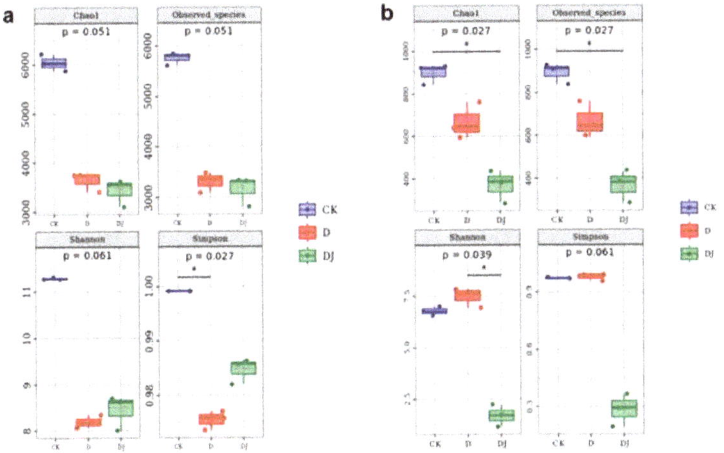

Figure 7. Plot of alpha diversity index of (**a**) bacteria and (**b**) fungi in soil samples supplemented with acetamiprid and dominant strains.

3.5.3. Beta Diversity Index Analysis

Beta diversity is represented by principal coordinates analysis (PCoA) to illustrate microbial community differences among different sample groups. The PCoA results (Figure 8) show that the three replicate samples of soil samples from the control group (CK) and two experimental groups (D and DJ) cluster in different quadrants. In bacterial samples (Figure 8a), the first principal component PCo1 (74.3%) has a greater impact on separation than the second principal component PCo2 (8.6%). The sampling points of the two experimental groups and the control group are distinctly separated, indicating significant changes in soil bacterial community composition influenced by acetamiprid and the dominant strain Md2. In fungal samples (Figure 8b), the first principal component PCo1 (52.3%) has a greater impact on separation than the second principal component PCo2 (27.8%). The distance between CK and DJ is significantly smaller than the distance between CK and D, suggesting that fungal community composition underwent significant changes with the invasion of acetamiprid and subsequent addition of Md2, which partially degraded the acetamiprid, resulting in a restoration of stability in fungal community composition over time. These results are further supported by the community composition at the phylum level.

are N-demethylation to form IM-2-1 and oxidative cleavage of the cyanoimino group to form IM-1-3. The toxicity of the metabolites was found to be lower than that of the parent compound.

Furthermore, the introduction of the acetamiprid-degrading dominant strain increased the relative abundance of certain phyla (such as *Firmicutes*, *Proteobacteria*, *Bacteroidetes*, and *Basidiomycota*) and genera (such as *Lysobacter*, *Acinetobacter*, and *Flavobacterium*) in the soil, possibly due to their involvement in the biodegradation of acetamiprid as a carbon and nitrogen source. This introduction also decreased the relative abundance of some other phyla (such as *Acidobacteria*, *Chloroflexi*, and *Ascomycota*) and genera (such as *Subgroup_6*, *Mortierellomycota*, *Fusarium*, and *Mycosphaerella*) in the soil. The dominant strain Md2 can effectively reduce acetamiprid residues in the environment, decrease the toxicity of the pesticide and its metabolites, and gradually adapt to contaminated soil, successfully colonizing it, thus contributing to the recovery of dominant phyla and genera in the soil microbiome.

Author Contributions: Conceptualization, X.W.; methodology, X.W. and M.Y.; software, L.G.; validation, J.B. and M.Y.; formal analysis, L.G.; resources, W.Y.; data curation, W.Y.; writing—original draft preparation, J.Z.; visualization, J.Z.; supervision, J.B. All authors have read and agreed to the published version of the manuscript.

Funding: This research was funded by the Applied Basic Research Plan of Liaoning Province (2023JH2/101300059).

Institutional Review Board Statement: Not applicable.

Informed Consent Statement: Not applicable.

Data Availability Statement: The data presented in this study are available on request from the corresponding author.

Acknowledgments: We are grateful for the diligent efforts of the editor and the anonymous reviewers. The authors also appreciate the teachers and students for their excellent help during the experiment. This article represents the views or opinions of its authors.

Conflicts of Interest: The authors declare no conflicts of interest.

References

1. Dolatabadi, M.; Naidu, H.; Ahmadzadeh, S. A green approach to remove acetamiprid insecticide using pistachio shell-based modified activated carbon; economical groundwater treatment. *J. Clean. Prod.* **2021**, *316*, 128226. [CrossRef]
2. Guo, W.; Yang, Y.; Zhou, X.; Ming, R.Y.; Hu, D.Y.; Lu, P. Insight into the toxic effects, bioconcentration and oxidative stress of acetamiprid on *Rana nigromaculata* tadpoles. *Chemosphere* **2022**, *305*, 135380. [CrossRef] [PubMed]
3. Xie, G.H.; Liu, G.G.; Dezhi, S.; Zheng, L.Q. Kinetics of Acetamiprid Photolysis in Solution. *Bull. Environ. Contam. Toxicol.* **2009**, *82*, 129–132. [CrossRef]
4. Roberts, T.; Hutson, D.; Lee, P.W.; Nicholls, P.H.; Plimmer, J.R. Metabolic pathways of agrochemicals, Part two: Insecticides and fungicide. In *Neonicotinoids*, 1st ed.; Roberts, T.R., Ed.; The Royal Society of Chemistry: London, UK, 1999; pp. 111–120. [CrossRef]
5. Wang, J.; Ohno, H.; Ide, Y.; Ichinose, H.; Mori, T.; Kawagishi, H.; Hirai, H. Identification of the cytochrome P450 involved in the degradation of neonicotinoid insecticide acetamiprid in Phanerochaete chrysosporium. *J. Hazard. Mater.* **2019**, *371*, 494–498. [CrossRef] [PubMed]
6. Cycoń, M.; Markowicz, A.; Borymski, S.; Wójcik, M.; Piotrowska-Seget, Z. Imidacloprid induces changes in the structure, genetic diversity and catabolic activity of soil microbial communities. *J. Environ. Manag.* **2013**, *131*, 55–65. [CrossRef] [PubMed]
7. Hallmann, C.A.; Foppen, R.P.B.; van Turnhout, C.A.M.; de Kroon, H.; Jongejans, E. Declines in insectivorous birds are associated with high neonicotinoid concentrations. *Nature* **2014**, *511*, 341–343. [CrossRef] [PubMed]
8. Mori, T.; Wang, J.; Tanaka, Y.; Nagai, K.; Kawagishi, H.; Hirai, H. Bioremediation of the neonicotinoid insecticide clothianidin by the white-rot fungus *Phanerochaete sordida*. *J. Hazard. Mater.* **2017**, *321*, 586–590. [CrossRef]
9. Kaur, R.P.; Gupta, V.; Christopher, A.F.; Bansal, P. Potential pathways of pesticide action on erectile function—A contributory factor in male infertility. *Asian Pac. J. Reprod.* **2015**, *4*, 322–330. [CrossRef]
10. Pimentel, D.; Levitan, L. Pesticides: Amounts Applied and Amounts Reaching Pests. *BioScience* **1986**, *36*, 86–91. [CrossRef]
11. Garcerá, C.; Moltó, E.; Chueca, P. Spray pesticide applications in Mediterranean citrus orchards: Canopy deposition and off-target losses. *Sci. Total Environ.* **2017**, *599–600*, 1344–1362. [CrossRef]

12. Jensen, P.D.; Sullivan, T.; Carney, C.; Batstone, D.J. Analysis of the potential to recover energy and nutrient resources from cattle slaughterhouses in Australia by employing anaerobic digestion. *Appl. Energy* **2014**, *136*, 23–31. [CrossRef]
13. Wang, F.; Yao, J.; Chen, H.; Yi, Z.; Choi, M.M. Influence of short-time imidacloprid and acetamiprid application on soil microbial metabolic activity and enzymatic activity. *Environ. Sci. Pollut. Res. Int.* **2014**, *21*, 10129–10138. [CrossRef] [PubMed]
14. Cycon, M.; Mrozik, A.; Piotrowska-Seget, Z. Antibiotics in the Soil Environment-Degradation and Their Impact on Microbial Activity and Diversity. *Front. Microbiol.* **2019**, *10*, 338. [CrossRef] [PubMed]
15. Wu, C.; Wang, Z.; Ma, Y.; Luo, J.; Gao, X.; Ning, J.; Mei, X.; She, D. Influence of the neonicotinoid insecticide thiamethoxam on soil bacterial community composition and metabolic function. *J. Hazard. Mater.* **2021**, *405*, 124275. [CrossRef] [PubMed]
16. Yao, X.-H.; Min, H.; Lu, Z.-H.; Yuan, H.-P. Influence of acetamiprid on soil enzymatic activities and respiration. *Eur. J. Soil Biol.* **2006**, *42*, 120–126. [CrossRef]
17. Hamada, A.; Wahl, G.D.; Nesterov, A.; Nakao, T.; Kawashima, M.; Banba, S. Differential metabolism of imidacloprid and dinotefuran by Bemisia tabaci CYP6CM1 variants. *Pestic. Biochem. Physiol.* **2019**, *159*, 27–33. [CrossRef] [PubMed]
18. Chen, S.; Chang, C.; Deng, Y.; An, S.; Dong, Y.H.; Zhou, J.; Hu, M.; Zhong, G.; Zhang, L.-H. Fenpropathrin Biodegradation Pathway in *Bacillus* sp. DG-02 and Its Potential for Bioremediation of Pyrethroid-Contaminated Soils. *J. Agric. Food Chem.* **2014**, *62*, 2147–2157. [CrossRef] [PubMed]
19. Rana, S.; Jindal, V.; Mandal, K.; Kaur, G.; Gupta, V.K. Thiamethoxam degradation by *Pseudomonas* and *Bacillus* strains isolated from agricultural soils. *Environ. Monit. Assess.* **2015**, *187*, 300. [CrossRef] [PubMed]
20. Xiao, Y.; Chen, S.; Gao, Y.; Hu, W.; Hu, M.; Zhong, G. Isolation of a novel beta-cypermethrin degrading strain *Bacillus subtilis* BSF01 and its biodegradation pathway. *Appl. Microbiol. Biotechnol.* **2015**, *99*, 2849–2859. [CrossRef]
21. Zhan, H.; Wang, H.; Liao, L.; Feng, Y.; Fan, X.; Zhang, L.; Chen, S. Kinetics and Novel Degradation Pathway of Permethrin in Acinetobacter baumannii ZH-14. *Front. Microbiol.* **2018**, *9*, 98. [CrossRef]
22. Bhatt, P.; Bhatt, K.; Huang, Y.; Lin, Z.; Chen, S. Esterase is a powerful tool for the biodegradation of pyrethroid insecticides. *Chemosphere* **2020**, *244*, 125507. [CrossRef]
23. Xu, B.; Xue, R.; Zhou, J.; Wen, X.; Shi, Z.; Chen, M.; Xin, F.; Zhang, W.; Dong, W.; Jiang, M. Characterization of Acetamiprid Biodegradation by the Microbial Consortium ACE-3 Enriched From Contaminated Soil. *Front. Microbiol.* **2020**, *11*, 1429. [CrossRef] [PubMed]
24. Zhai, Q.; Chen, X.; Zhang, M.; Zhang, C.; Zhang, Z.; Pan, H.; Zhang, H.; Sun, F. Immobilization of Klebsiella jilinsis strain 2N3 by corn straw biochar enhanced the degradation of nicosulfuron and restores the soil microbiome function and composition. *Appl. Soil Ecol.* **2023**, *189*, 104917. [CrossRef]
25. Zang, H.; Yu, Q.; Lv, T.; Cheng, Y.; Feng, L.; Cheng, X.; Li, C. Insights into the degradation of chlorimuron-ethyl by Stenotrophomonas maltophilia D310-3. *Chemosphere* **2016**, *144*, 176–184. [CrossRef] [PubMed]
26. Shi, Z.; Dong, W.; Xin, F.; Liu, J.; Zhou, X.; Xu, F.; Lv, Z.; Ma, J.; Zhang, W.; Fang, Y.; et al. Characteristics and metabolic pathway of acetamiprid biodegradation by Fusarium sp. strain CS-3 isolated from soil. *Biodegradation* **2018**, *29*, 593–603. [CrossRef]
27. Elango, D.; Siddharthan, N.; Alaqeel, S.I.; Subash, V.; Manikandan, V.; Almansour, A.I.; Kayalvizhi, N.; Jayanthi, P. Biodegradation of neonicotinoid insecticide acetamiprid by earthworm gut bacteria Brucella intermedium PDB13 and its ecotoxicity. *Microbiol. Res.* **2023**, *268*, 127278. [CrossRef] [PubMed]
28. Medhi, K.; Thakur, I.S. Bioremoval of nutrients from wastewater by a denitrifier Paracoccus denitrificans ISTOD1. *Bioresour. Technol. Rep.* **2018**, *1*, 56–60. [CrossRef]
29. Swati; Ghosh, P.; Thakur, I.S. Biodegradation of pyrene by *Pseudomonas* sp. ISTPY2 isolated from landfill soil: Process optimisation using Box-Behnken design model. *Bioresour. Technol. Rep.* **2019**, *8*, 100329. [CrossRef]
30. Zhang, X.; Huang, Y.; Chen, W.-J.; Wu, S.; Lei, Q.; Zhou, Z.; Zhang, W.; Mishra, S.; Bhatt, P.; Chen, S. Environmental occurrence, toxicity concerns, and biodegradation of neonicotinoid insecticides. *Environ. Res.* **2023**, *218*, 114953. [CrossRef]
31. Zamule, S.M.; Dupre, C.E.; Mendola, M.L.; Widmer, J.; Shebert, J.A.; Roote, C.E.; Das, P. Bioremediation potential of select bacterial species for the neonicotinoid insecticides, thiamethoxam and imidacloprid. *Ecotoxicol. Environ. Saf.* **2021**, *209*, 111814. [CrossRef]
32. Bhatti, S.; Satyanarayana, G.N.V.; Patel, D.K.; Satish, A. Bioaccumulation, biotransformation and toxic effect of fipronil in Escherichia coli. *Chemosphere* **2019**, *231*, 207–215. [CrossRef] [PubMed]
33. Wen-yi, L.; Rui-xia, F.; Peng-yu, Y.; Zhi, Y.; Yun-tao, G.; Hua-bin, X.; Xiao-fen, L.; Man-hong, L. Study on monitoring soil pesticide pollution by plant micronucleus technology. *J. Yunnan Minzu Univ. (Nat. Sci. Ed.)* **2021**, *30*, 532–535.
34. Papadopoulou, E.S.; Genitsaris, S.; Omirou, M.; Perruchon, C.; Stamatopoulou, A.; Ioannides, I.; Karpouzas, D.G. Bioaugmentation of thiabendazole-contaminated soils from a wastewater disposal site: Factors driving the efficacy of this strategy and the diversity of the indigenous soil bacterial community. *Environ. Pollut.* **2018**, *233*, 16–25. [CrossRef] [PubMed]
35. Li, H.; Qiu, Y.; Yao, T.; Ma, Y.; Zhang, H.; Yang, X.; Li, C. Evaluation of seven chemical pesticides by mixed microbial culture (PCS-1): Degradation ability, microbial community, and Medicago sativa phytotoxicity. *J. Hazard. Mater.* **2020**, *389*, 121834. [CrossRef] [PubMed]
36. Pelosi, C.; Joimel, S.; Makowski, D. Searching for a more sensitive earthworm species to be used in pesticide homologation tests—A meta-analysis. *Chemosphere* **2013**, *90*, 895–900. [CrossRef]

37. Huang, Y.C.; Xiao, L.J.; Li, F.Y.; Xiao, M.S.; Lin, D.R.; Long, X.M.; Wu, Z.J. Microbial Degradation of Pesticide Residues and an Emphasis on the Degradation of Cypermethrin and 3-phenoxy Benzoic Acid: A Review. *Molecules* **2018**, *23*, 2313. [CrossRef] [PubMed]
38. Iqbal, M. Vicia faba bioassay for environmental toxicity monitoring: A review. *Chemosphere* **2016**, *144*, 785–802. [CrossRef] [PubMed]
39. Hong-Zhi, Z. The Micronuclear Research Of The Environmental Pollution Caused By A Virulent Organic Phosphorus Pesticide (Parathion-Methyl). *J. Sichuan Univ. (Nat. Sci. Ed.)* **2000**, 62–66.
40. Ji, W.; Chen, T.; Sang, Q.; Dai, Y.; Ge, F.; Yuan, S. Metabolic Effects of Penicillium Oxalicum IM-3 on Chloropyridine Nicotinic Insecticides. *J. Ecol. Rural Environ.* **2010**, *26*, 246–250.
41. Hurek, T.; Wagner, B.; Reinhold-Hurek, B. Identification of N2-fixing plant- and fungus-associated Azoarcus species by PCR-based genomic fingerprints. *Appl. Environ. Microbiol.* **1997**, *63*, 4331–4339. [CrossRef]
42. Wang, C.-N.; Wu, R.-L.; Li, Y.-Y.; Qin, Y.-F.; Li, Y.-L.; Meng, F.-Q.; Wang, L.-G.; Xu, F.-L. Effects of pesticide residues on bacterial community diversity and structure in typical greenhouse soils with increasing cultivation years in Northern China. *Sci. Total Environ.* **2020**, *710*, 136321. [CrossRef] [PubMed]
43. Pan, X.; Wang, S.; Shi, N.; Fang, H.; Yu, Y. Biodegradation and detoxification of chlorimuron-ethyl by *Enterobacter ludwigii* sp. CE-1. *Ecotoxicol. Environ. Saf.* **2018**, *150*, 34–39. [CrossRef] [PubMed]
44. Pang, S.M.; Lin, Z.Q.; Zhang, W.P.; Mishra, S.; Bhatt, P.; Chen, S.H. Insights Into the Microbial Degradation and Biochemical Mechanisms of Neonicotinoids. *Front. Microbiol.* **2020**, *11*, 00868. [CrossRef] [PubMed]
45. Zhang, Y.; Zhang, X.L.; Zhang, H.W.; He, Q.; Zhou, Q.X.; Su, Z.C.; Zhang, C.G. Responses of Soil Bacteria to Long-Term and Short-Term Cadmium Stress as Revealed by Microbial Community Analysis. *Bull. Environ. Contam. Toxicol.* **2009**, *82*, 367–372. [CrossRef] [PubMed]
46. Desai, C.; Parikh, R.Y.; Vaishnav, T.; Shouche, Y.S.; Madamwar, D. Tracking the influence of long-term chromium pollution on soil bacterial community structures by comparative analyses of 16S rRNA gene phylotypes. *Res. Microbiol.* **2009**, *160*, 1–9. [CrossRef]
47. Huang, L.-N.; Zhu, S.; Zhou, H.; Qu, L.-H. Molecular phylogenetic diversity of bacteria associated with the leachate of a closed municipal solid waste landfill. *FEMS Microbiol. Lett.* **2005**, *242*, 297–303. [CrossRef]
48. Sánchez, O.; Ferrera, I.; González, J.M.; Mas, J. Assessing bacterial diversity in a seawater-processing wastewater treatment plant by 454-pyrosequencing of the 16S rRNA and amoA genes. *Microb. Biotechnol.* **2013**, *6*, 435–442. [CrossRef] [PubMed]
49. Yu, B.; Chen, Z.; Lu, X.; Huang, Y.; Zhou, Y.; Zhang, Q.; Wang, D.; Li, J. Effects on soil microbial community after exposure to neonicotinoid insecticides thiamethoxam and dinotefuran. *Sci. Total Environ.* **2020**, *725*, 138328. [CrossRef]
50. Liang, Y.M.; Pan, F.J.; Ma, J.M.; Yang, Z.Q.; Yan, P.D. Long-term forest restoration influences succession patterns of soil bacterial communities. *Environ. Sci. Pollut. Res.* **2021**, *28*, 20598–20607. [CrossRef]
51. Yan, B.S.; Sun, L.P.; Li, J.J.; Liang, C.Q.; Wei, F.R.; Xue, S.; Wang, G.L. Change in composition and potential functional genes of soil bacterial and fungal communities with secondary succession in *Quercus liaotwigensis* forests of the Loess Plateau, western China. *Geoderma* **2020**, *364*, 114199. [CrossRef]
52. Nettles, R.; Watkins, J.; Ricks, K.; Boyer, M.; Licht, M.; Atwood, L.W.; Peoples, M.; Smith, R.G.; Mortensen, D.A.; Koide, R.T. Influence of pesticide seed treatments on rhizosphere fungal and bacterial communities and leaf fungal endophyte communities in maize and soybean. *Appl. Soil Ecol.* **2016**, *102*, 61–69. [CrossRef]
53. Zabaloy, M.C.; Gómez, E.; Garland, J.L.; Gómez, M.A. Assessment of microbial community function and structure in soil microcosms exposed to glyphosate. *Appl. Soil Ecol.* **2012**, *61*, 333–339. [CrossRef]
54. Upadhyay, L.S.B.; Dutt, A. Microbial Detoxification of Residual Organophosphate Pesticides in Agricultural Practices. In *Microbial Biotechnology*; Springer: Singapore, 2017; pp. 225–242.
55. Horemans, B.; Vandermaesen, J.; Vanhaecke, L.; Smolders, E.; Springael, D. Variovorax sp.-mediated biodegradation of the phenyl urea herbicide linuron at micropollutant concentrations and effects of natural dissolved organic matter as supplementary carbon source. *Appl. Microbiol. Biotechnol.* **2013**, *97*, 9837–9846. [CrossRef] [PubMed]
56. Mahapatra, B.; Adak, T.; Patil, N.K.B.; Gowda, G.B.; Jambhulkar, N.N.; Yadav, M.K.; Panneerselvam, P.; Kumar, U.; Munda, S.; Jena, M. Imidacloprid application changes microbial dynamics and enzymes in rice soil. *Ecotoxicol. Environ. Saf.* **2017**, *144*, 123–130. [CrossRef]
57. Villaverde, J.; Rubio-Bellido, M.; Merchán, F.; Morillo, E. Bioremediation of diuron contaminated soils by a novel degrading microbial consortium. *J. Environ. Manag.* **2017**, *188*, 379–386. [CrossRef]

toxics

MDPI

Article

Interaction of Microplastics with Emerging Organic Pollutants: A Study on Atrazine Adsorption and Phytotoxicity

Luan Gabriel Xavier de Souza [1], Francisco Javier Cuba Teran [1,*], Renata Medici Frayne Cuba [1], Andréa Rodrigues Chaves [2] and Kellen Cristina da Silva [1]

[1] Civil and Environmental School, Federal University of Goiás, Goiania 74605-220, Brazil; luan.gabriel@ufg.br (L.G.X.d.S.); renatafrayne@ufg.br (R.M.F.C.); kellen_cristina@discente.ufg.br (K.C.d.S.)
[2] Chemistry Institute, Federal University of Goiás, Goiania 74001-970, Brazil; andrea_chaves@ufg.br
* Correspondence: paco@ufg.br

Abstract: The adsorption of atrazine (ATZ) onto pristine and aged polyethylene microplastics (MPs) was investigated in distilled water (DW) and hydroponic nutrient-enriched water (EW) to evaluate its phytotoxic effects on *Lactuca sativa* germination. Aged microplastics (AMPs) exhibited higher ATZ adsorption in both conditions: 0.646 mg/g (14.49%) in DW and 0.742 mg/g (15.87%) in EW, compared to 0.405 mg/g (9.08%) and 0.504 mg/g (10.78%) for pristine microplastics (PMPs), respectively. This increase was attributed to photodegradation-induced surface modifications on MP, including increased roughness and the formation of oxygenated functional groups. The phytotoxicity assays showed that ATZ adsorbed onto AMPs inhibited seed germination more severely, with a maximum inhibition of 34% at 2 mg/L, evidencing that microplastic aging enhances ATZ adsorption and increases toxicity risks in aquatic environments, particularly under eutrophic conditions. The combined presence of MP and ATZ resulted in greater toxicity, attributed to a synergistic effect, as observed in dry and wet mass inhibition. These findings indicate that pollutant interactions amplify negative impacts on plant development. Furthermore, ATZ primarily affects root growth through direct physical contact with MP rather than via desorption into water.

Keywords: microplastic; phytotoxicity; atrazine

1. Introduction

Plastic pollution is an escalating global issue affecting various environmental compartments, including soil, water, and air [1]. The global production of approximately 430 million tons of plastic annually raises serious concerns [2]. A significant concern is the fragmentation of plastics into MPs, a process occurring both naturally, through macroplastic degradation, and intentionally via industrial production for applications in construction, cosmetics, and nautical industries [3].

MP particles, typically ranging from 1 to 5 mm in size, have been extensively studied in the context of plastic pollution due to their adverse effects. These effects range from developmental modifications in photosynthetic organisms—leading to oxidative stress, as observed in phytotoxicity tests [4]—to impacts on top-tier organisms, particularly marine species, where MP ingestion frequently results in gastrointestinal obstruction and suffocation [5].

Beyond their direct effects, MPs can modify pollutant toxicity and bioavailability [6,7]. In aquatic environments, these particles interact with organic and inorganic pollutants, promoting contaminant retention through adsorption processes [8]. Under specific environmental conditions like weathering, salinity, pH, and heavy metals, these pollutants may transfer onto MPs and later be released through desorption mechanisms [9–11]. Studies have demonstrated that MPs can adsorb pesticides, including ATZ, affecting their environmental fate and potential bioavailability [12,13]. Research indicates that MPs enhance the persistence of ATZ in aquatic environments, increasing its potential for bioaccumulation and ecological risks [14]. Furthermore, microbial aging and environmental factors alter MP's surface characteristics, modifying ATZ adsorption behaviors [15].

Investigating these interactions is crucial due to their potential to exacerbate environmental pollution, particularly through interactions with widely used organic contaminants such as pesticides. With global population growth, the demand for agricultural pesticides increases proportionally to ensure food supply [16]. The widespread use of these substances in agriculture raises concerns regarding their environmental impacts [17], particularly their growing presence in soils [18] and aquatic compartments [19]. Among the numerous pesticides, some are of particular concern due to their extensive use, prolonged persistence, long residual time, and stability—such as ATZ [20].

ATZ is extensively used and is among the most consumed pesticides in Brazil [21] and globally. AZT is applied as a pre- and post-emergent herbicide in agricultural practices, particularly in maize, sugarcane, and sorghum cultivation [22]. Classified as a highly hazardous pesticide by the Pesticide Action Network (PAN), ATZ contamination in natural environments, including irrigation water and agricultural drainage [23,24], represents a significant concern. Due to its moderate to high persistence in soil and aquatic environments, ATZ can remain in ecosystems for weeks to months, particularly in groundwater, where degradation is slower, leading to long-term exposure risks [25]. Moreover, it has been associated with adverse health effects, including potential endocrine disruption and reproductive toxicity [26,27]. The interaction between MPs and ATZ is an emerging environmental issue, as such interactions are becoming increasingly frequent, potentially leading to combined contamination [28].

Previous studies [29–32] have primarily focused on MP–ATZ interactions in single-phase systems. However, the influence of environmentally relevant water compositions, particularly under nutrient-rich conditions, remains underexplored. Additionally, while many studies have examined MP and pesticide interactions from a physicochemical perspective, the combined effects of these contaminants on plant health and phytotoxicity have received limited attention. By assessing MP–ATZ interactions in both DW and EW and incorporating phytotoxicity evaluations, our study provides a more comprehensive understanding of how these pollutants behave in real aquatic environments and their potential ecological risks.

The effects of ATZ on the environment extend beyond target organisms, disrupting the life cycle of various species due to water and soil contamination [31]. ATZ undergoes chemical transformations that enhance its conversion into toxic and persistent compounds [32]. Its toxicity has been reported in humans (disrupting hypothalamic control and affecting primary hepatocytes and hepatic stem cells), as well as in non-target plants and microbes (inducing oxidative stress and reducing population size) [33].

Thus, the objective of this research is to investigate, through adsorption and desorption experiments, the interaction between ATZ and MPs. This study aims to evaluate this interaction in different matrices, including DW and hydroponic EW. Additionally, the phytotoxic effects of this interaction on *Lactuca sativa* will be assessed through phytotoxicity assays.

2. Materials and Methods

2.1. Simulation of Environmental Conditions

This study was conducted under simulated environmental conditions, using DW (pH 6) and hydroponic EW (pH 7), formulated with salts and minerals commonly found in hydroponic systems. These matrices were employed to explore the interactions between MPs and contaminants in different representative scenarios to evaluate the influence of water composition on the interactions between PMPs, AMPs, and ATZ.

2.2. Microplastic Acquisition and Photodegradation

Polyethylene microspheres, with an average diameter of 0.5 mm, were purchased from Bianquimica, São Paulo, Brazil. The samples underwent photodegradation in an accelerated aging chamber for 120 days. Continuous UV radiation was applied, maintaining a controlled average temperature of 32 °C. To ensure uniform exposure to radiation, weekly redistribution of the microspheres within the chamber was performed. These conditions were selected based on their relevance in simulating environmental aging processes and assessing the photodegradation behavior of contaminants. The use of an accelerated aging chamber with continuous UV radiation and a controlled temperature of 32 °C aligns with established methodologies for evaluating the stability and transformation of pollutants under realistic yet controlled conditions [34,35].

Previous studies have demonstrated that the fragmentation of micro- and nanoplastics, induced by photo-oxidation in accelerated aging chambers, varies among different polymers, highlighting the importance of simulating environmental conditions to understand these processes [35]. The 120-day exposure period was chosen to capture potential degradation pathways over extended exposure, simulating long-term environmental interactions. These parameters maintain consistency with previous studies, facilitating comparison and validation of the observed results.

2.3. Microplastic Characterization

MPs were characterized based on their morphological and physicochemical properties. Morphological analysis was conducted using scanning electron microscopy (SEM) (Jeol, JSM-6610 Tokyo, Japan), while the physicochemical characterization included zeta potential (ZP) measurements (Zetasizer Nano Series ZS, Marvel, Worcestershire, UK) and Fourier transform infrared spectroscopy (FTIR) to assess the surface modifications in MPs (Bruker Vertex 70, Ettingen, Germany). The FTIR measurements were conducted with a resolution of 4 cm^{-1}, using 32 scans and an acquisition range of 4000–400 cm^{-1}. Prior to analysis, the samples were pre-filtered using a 0.4 µm syringe filter (PTFE MerckMillipore, Darmstadt Hesse, Germany) to remove impurities and undesirable particles. The filters were pre-saturated before use, and the initial filtrate was discarded.

The Carbonyl Index (CI) was calculated to evaluate MP aging, determined as the ratio of the carbonyl absorption band area (1850–1650 cm^{-1}) to the reference methylene band area (1500–1420 cm^{-1}), as shown in Equation (1).

$$CI = \frac{\text{Carbonyl band area } (1850 - 1650 \text{ cm}^{-1})}{\text{Reference band area } (1500 - 1420 \text{ cm}^{-1})} \quad (1)$$

2.4. Adsorption and Desorption Assays

Preliminary assays were conducted to determine the optimal concentration of polyethylene microspheres based on their adsorption capacity. A 2 mg/L ATZ solution was used, varying the MP mass from 5 mg to 30 mg in 50 mL of ATZ solution. The samples were continuously agitated (120 rpm) for 48 h. The removal efficiencies and adsorption capacities were calculated using Equation (2).

$$q_t = \frac{(C_0 - C_t)V}{m} \quad (2)$$

where q_t is the adsorption capacity, C_0 and C_t represent the initial and final concentrations of ATZ, V is the solution volume, and m is the mass of microplastic.

To determine the adsorption kinetics, the effect of the contact time was evaluated at intervals of 0, 15, 30, 60, 240, 1440, and 2880 min, enabling data modeling using pseudo-first-order (PFO) and pseudo-second-order (PSO) models. The experiments were conducted using 0.025 g of MPs (a dosage considered optimal based on preliminary experiments) in 50 mL of a 2 mg/L ATZ solution, maintained under constant agitation (120 rpm) until adsorption equilibrium was reached.

Following the determination of the optimal MP dosage in the preliminary test, adsorption isotherms were obtained by varying the initial ATZ concentration at 0.05, 0.1, 0.5, 1.0, 2.0, and 4.0 mg/L while keeping the MP mass constant. The experiments were conducted at 120 rpm for 48 h to ensure the stabilization of interactions between the contaminant and the adsorbent. The obtained data were fitted to the Langmuir and Freundlich models to determine the maximum adsorption capacity (Langmuir) and adsorption intensity (Freundlich).

For the desorption experiments, MPs that reached adsorption equilibrium under conditions of 2 mg/L ATZ and 0.025 g of MP were filtered and transferred to clean solutions, where they remained for 48 h under the same experimental conditions (120 rpm, 24 °C). The amount of released ATZ was monitored to assess the reversibility of adsorption and understand its environmental implications.

2.5. Phytotoxicity Assays

Phytotoxicity assays were conducted using *Lactuca sativa* seeds to evaluate the impact of MPs and ATZ on germination and root growth processes. Four treatment types were tested as follows: a negative control consisting of DW and EW without MPs or ATZ; a positive control with 0.05 mol/L of zinc sulfate ($ZnSO_4$) solution [36–38], used as a reference for known toxicity to validate the test sensitivity; treatments with isolated MPs, where seeds were exposed to MPs in distilled or enriched water without ATZ; and treatments with MPs containing adsorbed ATZ (MP+ATZ), in which seeds were exposed to MPs pre-treated with ATZ in both aquatic matrices.

The seeds were placed in Petri dishes containing filter paper moistened with 4 mL of the corresponding solution for each treatment. The plates were incubated at a controlled temperature of 24.5 °C for 120 h in the dark. The number of MPs in each treatment was determined based on the highest ATZ adsorption capacity previously observed (0.025 g) and was conducted in triplicate. For the treatment with ATZ alone, the concentration from the adsorption assays (2 mg/L) was used. After incubation, the treatment effects were evaluated using quantitative indicators, including the Root Growth Index (ICR), Germination Index (IG), Normalized Root Growth Percentage Index (IER), and Normalized Residual Germination Percentage Index (IGN) [39].

Additionally, the inhibition percentages of the fresh and dry biomasses were assessed [4] using the obtained data. The indices were calculated using Equations (3)–(7).

Root Growth Index (ICR):

$$ICR = \frac{CRA}{CRC} \quad (3)$$

Germination Index (IG):

$$IG = ICR \times \left(\frac{SGA}{SGC}\right) \times 100 \quad (4)$$

Normalized Root Growth Percentage Index (IER):

$$IER(\%) = \frac{CMRCA - CMRCN}{CMRCN} \quad (5)$$

Normalized Residual Germination Percentage Index (IGN):

$$IGN(\%) = \frac{GERM\ sample - GERM\ control}{GERM\ control} \quad (6)$$

Inhibition of Dry and Fresh Biomass (INM) (%):

$$\frac{Mass\ in\ Control - Mass\ in\ Treatment}{(Mass\ in\ Control)} \times 100 \quad (7)$$

Variable Definitions:
ICR: Root Growth Index.
CRA: Root Growth in Sample.
CRC: Root Growth in the Control.
IG: Germination Index.
SGA: Number of Germinated Seeds in Sample.
SGC: Number of Germinated Seeds in Negative Control.
IER: Normalized Root Growth Percentage Index
CMRCA: Average Radicle Length in Each Sample
CMRCN: Average Radicle Length of Germinated Seeds in the Control.
IGN: Normalized Residual Germination Percentage Index.
GERMsample: Average Percentage of Germinated Seeds in Sample.
GERMcontrol: Percentage of Germinated Seeds in Negative Control.

These calculations followed the methodologies described in previous studies [40]. The NRGPI and NRGPI values were used to classify the toxicity levels according to Table 1.

Table 1. Toxicity classification [40].

IGN (%) and IER (%)	Toxicity
0 to −0.25	Low
−0.25 to −0.5	Moderate
−0.5 ato−0.75	High
−0.75 ato −1.0	Very high

All treatments were conducted in triplicate, and the error bars presented in the graphs represent the standard deviation of the means (mean ± SD). Standard deviation values were calculated using the statistical software packages RStudio (version 2024,12,1-563) and Microsoft Excel (Microsoft Office Home and Business 2016). Prior to statistical analyses, the Shapiro–Wilk normality test ($\alpha = 0.05$) was performed to assess the normality of the residuals. When normality was confirmed, and ANOVA indicated significance ($p \leq 0.05$), Tukey's multiple comparisons test was applied to identify significant differences between treatments. For data that did not meet the normality criteria, the non-parametric Kruskal–Wallis test was used, followed by Dunn's multiple comparisons test.

3. Results

3.1. Surface Analysis of Microplastic by SEM and FTIR

SEM revealed significant structural changes in AMPs (Figure 1A,B). While the PMPs exhibited smooth and homogeneous surfaces (Figure 1C,D), the AMPs displayed roughness, cracks, and surface cavities, characteristics indicative of structural degradation.

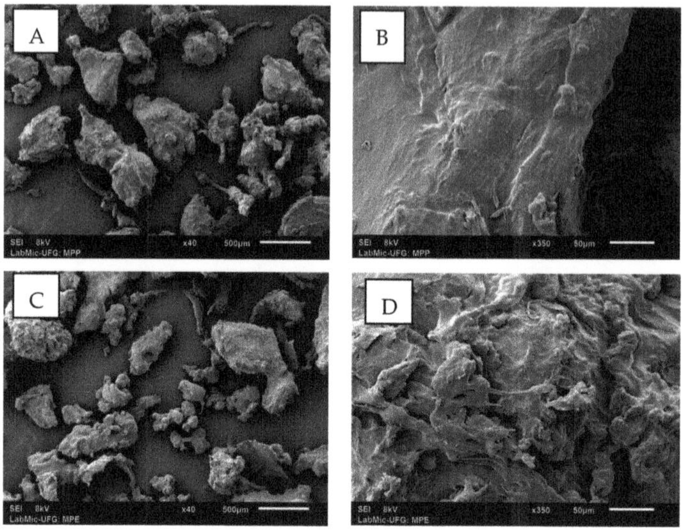

Figure 1. Image (**A**): SEM of PMPs (40×), displaying irregular morphology and rough edges, scale bar: 500 μm. Image (**B**): SEM of PMPs (350×), scale bar: 50 μm, exhibiting a smooth surface with no deformations. Image (**C**): SEM of AMPs (40×), scale bar: 500 μm, with fragmented particles and irregular distribution (scale bar: 500 μm). Image (**D**): AMPs (350×), showing a worn texture and thin layers (scale bar: 50 μm), scale bar: 50 μm.

FTIR was employed to assess the changes in the functional groups of MP particles before and after ATZ adsorption, as well as the structural modifications in the MPs due to aging (Figure 2). The spectral peaks observed for the PE MPs were attributed to -CH_2- stretching at 2915 cm^{-1} and 2848 cm^{-1}, -CH_2- bending at 1464 cm^{-1}, and -(CH_2)n- rocking at 719 cm^{-1}. After adsorption, spectral peaks were identified at 3254 cm^{-1} (-NH stretching) and 1550 cm^{-1} (thiotriazine stretching), which are characteristic of ATZ but appeared with low intensity, suggesting that the interaction between this compound and both PMPs and AMPs was primarily physical [15]. Aging of polyethylene MPs due to UV radiation exposure for 120 days resulted in significant chemical changes on their surfaces, as observed through the Fourier transform infrared (FTIR) spectroscopy analyses. These alterations are primarily evident in the spectral regions corresponding to the surface oxygenated groups, such as hydroxyl groups (3600–3000 cm^{-1}), carbonyls (1800–1400 cm^{-1}), and esters and vinyl groups (1400–800 cm^{-1}), as reported in previous studies [41,42].

One of the main spectral modifications observed in the FTIR spectrum was within the 1800–1500 cm^{-1} range, where carbonyl (C=O) peak reshaping occurred, with a broad peak at 1713 cm^{-1}, characterizing the formation of carbonyl groups due to UV-induced oxidation [43,44]. This peak was more pronounced in AMPs compared to the PMPs, as also reported by Zidar et al. [45]. Additionally, in the 3600–3000 cm^{-1} region, short vibration remodeling was observed in a broad band associated with the presence of hydroxyl (-OH) groups.

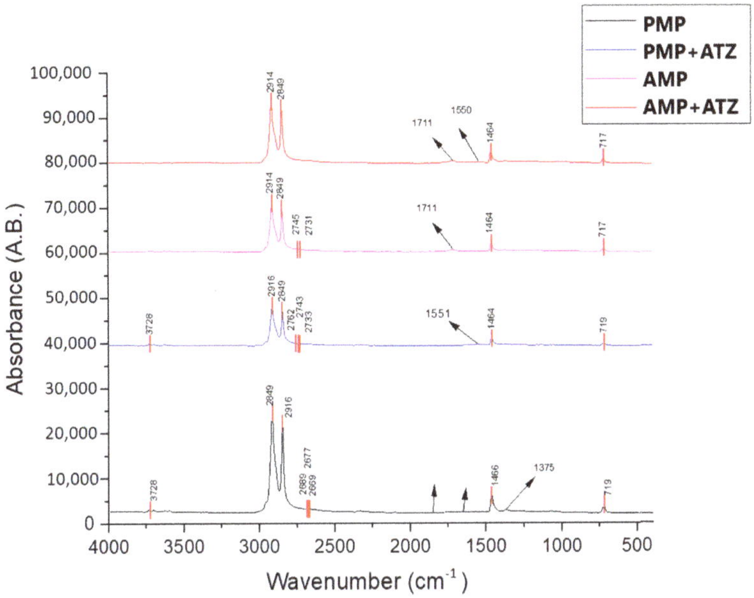

Figure 2. FTIR spectra of PE plastic fragments: The AMPs are shown in red, and the control PMPs are shown in black.

To quantify these structural modifications, the Carbonyl Index (CI) was calculated as the ratio of the carbonyl absorption band area (1850–1650 cm^{-1}) to the reference methylene band area (1500–1420 cm^{-1}). In PMPs, the CI was determined to be 1.623, whereas in AMPs, it reached 2.357, representing a significant 45.2% increase in carbonyl group formation due to aging. The rise in the CI further supports the chemical degradation of polymer chains, confirming significant alterations in the material's surface properties upon aging [46].

These results demonstrate that aging not only promoted the formation of carbonyl groups but also introduced other oxygenated functional groups, such as hydroxyls and carbonyls, increasing surface polarity. This photodegradation process altered both the chemical and physical properties of the MPs, making them more reactive and potentially more susceptible to interactions with polar environmental compounds such as ATZ.

The zeta potential measurements (Figure 3) further confirmed these changes, showing more negative values in AMPs (−32 mV) compared to PMPs (−26 mV), reinforcing the higher capacity of AMPs to interact with polar molecules [18,20]. These structural and chemical modifications highlight the impact of aging on the reactivity and environmental behavior of MPs [47], showing a greater surface density of functional groups, such as carbonyl and hydroxyl, which are introduced during the aging process due to oxidative weathering. In contrast, PMPs, with a less negative zeta potential, possessed a relatively lower surface reactivity, limiting its ability to engage in similar interactions.

Even considering the negatively charged microplastic, ATZ is a neutral to weakly basic herbicide with a pKa of 1.68, meaning it remains predominantly neutral at environmental pH levels (e.g., pH 5–8). However, its adsorption onto aged MPs can modify the surface properties by introducing hydrophobic patches or altering the charge distribution, thereby reducing overall surface negativity or even creating localized positive sites. Consequently, MPs with adsorbed ATZ can engage in electrostatic interactions with plant roots, with the nature of this interaction depending on the surface charge modifications induced by ATZ adsorption, as we see in Figure 3.

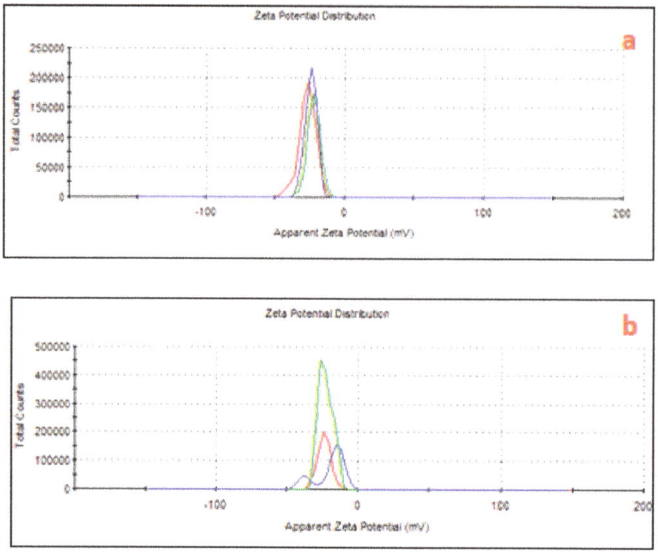

Figure 3. Zeta potential of AMPs (**a**) and PMPs (**b**), with adsorbed ATZ. The line colors refer to three replicates of the same sample.

3.2. Adsorption Capacity

The analysis of ATZ adsorption on PMPs and AMPs, considering the influence of the surrounding medium, reveals important insights into the adsorption mechanisms and the factors that influence them.

The obtained data indicate that both the properties of the MPs and the characteristics of the medium play significant roles in sorption efficiency. These variables should be evaluated together for a comprehensive understanding of contaminant behavior in aquatic environments. In the adsorption capacity assay, the results indicated that adsorption was most efficient with 25 mg of microplastics. This mass was selected due to its higher qe value compared to the other tested quantities.

When comparing AMPs with PMPs, a slight difference in ATZ adsorption capacity is observed. In EW, the AMPs adsorbed a slightly higher amount of ATZ (0.742 mg/g) compared to the PMPs (0.5042 mg/g), as clearly depicted in Figure 4. This behavior may be attributed to the chemical modifications occurring on the MP's surface during the aging process. The FTIR analysis suggested the emergence of carboxyl (C=O) groups due to the oxidation of PE particles. These functional groups may interact with the ethylamine moiety of ATZ molecules, which acts as a hydrogen bond donor [48].

Although the data indicate a slight advantage for AMPs in terms of ATZ sorption efficiency, the observed differences are relatively small. This suggests that the affinity of ATZ for both types of MPs in DW and EW media is low. Despite the modest absolute amount of ATZ adsorbed by the MPs (0.3 mg/L), it significantly exceeds the maximum allowable limits in drinking water, such as the 0.1 µg/L threshold established by the European Union [49].

As shown in Figure 4, MPs in the enriched medium exhibited higher adsorption values for AMPs and PMPs. This phenomenon may be related to the presence of ions, which influence the adsorption process through mechanisms such as electrostatic shielding, salting-out effects, and extrusion. The presence of NaCl, for example, tends to induce a salting-out effect on the adsorption of hydrophobic chemical compounds [50], which may have influenced the results observed in this study, given the presence of this salt in the enriched solutions.

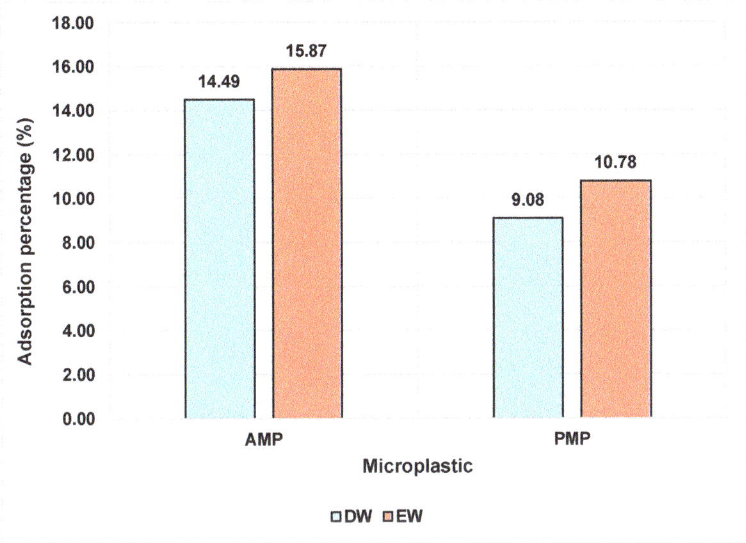

Figure 4. Sorption efficiency at equilibrium for ATZ and PMPs in DW and EW media. Experimental conditions: temperature = 25 °C; ATZ concentration = 2.0 mg L^{-1}; contact time = 24 h.

Furthermore, the presence of certain inorganic ions, beyond sodium and chloride, in natural water bodies can also influence adsorption. Studies have demonstrated that ions such as Cl^-, $SO_4{}^{2-}$, and $HCO_3{}^-$ enhance ATZ adsorption capacity to varying degrees, while ions such as Mg^{2+} and Ca^{2+} may inhibit this process. For instance, in the presence of Mg^{2+}, the adsorption capacity of PMPs decreased by 25.9%, whereas AMPs exhibited only a 4% reduction. This suggests that electrostatic attraction has a limited effect [18].

This phenomenon occurs mainly because ions such as Mg^{2+} and Ca^{2+} compete for adsorption sites on MPs. Being positively charged, they can be adsorbed onto the particles via electrostatic attraction, occupying the same adsorption sites as ATZ [51].

3.3. Adsorption Kinetics

The adsorption kinetic results were evaluated using pseudo-first-order and pseudo-second-order models, with the respective parameters presented in Table 2 and Figure 5. The analysis revealed that equilibrium was reached after approximately 48 h, with a sharp increase in adsorption rates during the initial stages of the experiment. The qe values (adsorption capacity at equilibrium) varied among treatments, with AMPs exhibiting slightly higher adsorption capacities compared to PMPs in both matrices (DW and EW).

Table 2. Kinetic adsorption parameters of ATZ onto PMPs and AMPs under different aquatic matrices: DW and EW.

Matrix	Control	AV	Pseudo-First-Order			Pseudo-Second-Order		
			q_e	k_1	R^2	q_e	k_2	R^2
DW	ATZ	PMP	0.427	0.07	0.9773	0.440	0.37	0.9833
DW	ATZ	AMP	0.505	0.100	0.9680	0.522	0.42	0.9877
EW	ATZ	PMP	0.566	0.197	0.9985	0.572	1.75	0.9998
EW	ATZ	AMP	0.603	0.06	0.9758	0.623	0.18	0.9869

q_e (mg g^{-1}); k_1 (h^{-1}); k_2 (g mg^{-1} h^{-1}).

A B

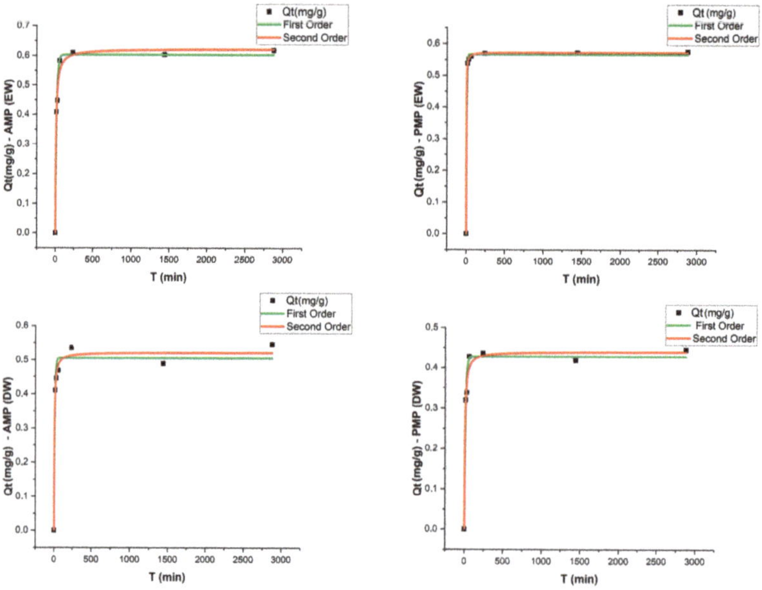

C D

Figure 5. Sorption kinetics of ATZ on (**A**) AMPs in EW, (**B**) PMPs in EW, (**C**) AMPs in DW, and (**D**) PMPs in DW. Experimental conditions: temperature = 25 °C; polyethylene concentration = 0.5 g L^{-1}.

For the pseudo-first-order model, the qe values for PMPs were 0.427 mg/g in DW and 0.566 mg/g in EW. In contrast, for AMPs, these values increased to 0.505 mg/g and 0.603 mg/g, respectively. The determination coefficients (R^2) for this model ranged from 0.968 to 0.9985, indicating a good fit of the experimental data to the model under both conditions.

In the pseudo-second-order model, the qe values were slightly higher, reaching 0.522 mg/g for AMPs in DW and 0.623 mg/g in EW. These values suggest that the adsorption mechanism is influenced by stronger chemical interactions between ATZ and AMPs. The k_2 coefficients for both PMPs and AMPs were higher in EW, with PMPs showing a particularly high k_2 value (1.75) and an excellent fit (R^2 = 0.9998).

When comparing the two models, the pseudo-second-order model exhibited higher R^2 coefficients (ranging from 0.9833 to 0.9998), suggesting that this model better describes the adsorption process of ATZ onto MPs, particularly in EW.

Figure 5 presents the experimental data along with the model fittings. These results highlight the influence of aging and the enriched aqueous matrix on the adsorption kinetics, demonstrating that the higher polarity of AMPs enhance ATZ retention.

3.4. Isotherms

The Langmuir and Freundlich isotherm models were applied to describe the adsorption of ATZ onto both PMPs and AMPs in two aquatic matrices: DW and EW. The Langmuir model provided the best fit for most treatments, with the R^2 values exceeding 0.9, indicating good agreement with the experimental data. This result suggests that adsorption occurs in monolayers with a homogeneous adsorption surface (Table 3, Figure 6).

Table 3. Isothermal adsorption parameters of ATZ onto PMPs and AMPs under different aquatic matrices: DW and EW.

Matrix	Control	AV	Langmuir			Freundlich		
			K_L	q_m	R^2	K_F	n	$Adj.\ R^2$
DW	ATZ	PMP	0.145	1.612	0.997	0.196	1.201	0.991
DW	ATZ	AMP	0.671	0.841	0.975	0.300	1.679	0.922
EW	ATZ	PMP	0.204	1.493	0.953	0.239	1.272	0.934
EW	ATZ	AMP	0.495	1.076	0.948	0.321	1.584	0.901

K_L (L/mg); q_m (mg/g); K_F (L/g); n adimensional.

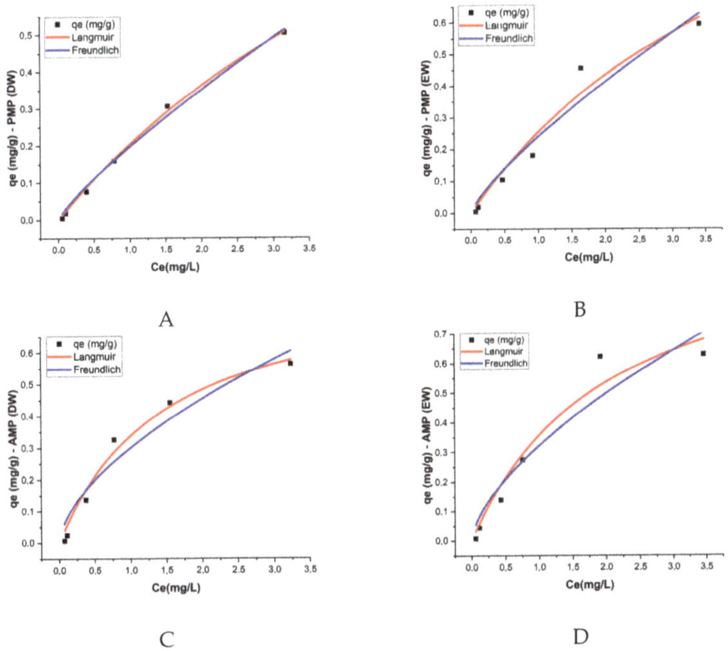

Figure 6. Sorption isotherms for (**A**) AMPs in EW, (**B**) AMPs in EW, (**C**) AMPs in DW, and (**D**) PMPs in DW. Experimental Conditions: temperature = 25 °C; polyethylene concentration = 0.5 g/L.

In PMPs in DW, the maximum adsorption capacity (qm) was 1.612 mg/g, while the Langmuir constant (KL) was 0.145 L/mg, reflecting a moderate affinity between ATZ and the PMP surfaces. For AMPs in DW, the maximum adsorption capacity decreased to 0.841 mg/g, but the K_L constant significantly increased to 0.671 L/mg, indicating that despite the reduction in total adsorption capacity, the affinity between AMPs and ATZ was higher. In EW, PMPs exhibited a qm of 1.493 mg/g and a KL of 0.204 L/mg, while AMPs had a lower qm of 1.076 mg/g but a higher KL of 0.495 L/mg.

Wang et al. [52] reported qe values for ATZ adsorption onto different types of MPs that are of the same order of magnitude as those obtained in the present study, specifically 1.349 mg/g, 1.992 mg/g, and 1.148 mg/g. After aging, the authors observed increases of 84%, 78%, and 95% for polystyrene, polyethylene, and polypropylene, respectively. In contrast, our study reported a decrease in qm by 52% in DW and 74% in EW after aging. Prolonged UV exposure, as applied in our study, may lead to more extensive polymer degradation, fragmentation, and morphological alterations, such as reduced surface area or pore collapse, which could negatively impact adsorption capacity. When comparing the qm and K_L values obtained by Wang et al. [52] with those in our study, the increase in K_L values for AMPs in both cases suggests that differences in adsorption behavior may be attributed to distinct morphological changes resulting from variations in the aging duration. In the present study, MPs were exposed to UV radiation for 120 days, whereas Wang et al. [52] applied UV exposure for only 96 h. According to Chen et al. [53], prolonged UV exposure induces more pronounced alterations in the surface structure of MPs. This is further supported by Zhang et al. [54], who reported a reduction in the surface area of three types of MPs after 30 days of UV aging. The Freundlich model exhibited a slightly lower R^2 compared to the Langmuir model, ranging from 0.901 to 0.991, yet still demonstrated good agreement with the experimental data, suggesting that the process occurs on heterogeneous surfaces with non-uniform energetic interactions. The values of the K_F constant, which reflects the adsorption capacity, were higher for AMPs in both matrices, with $K_F = 0.3$ in DW and $K_F = 0.321$ in EW (Table 3; Figure 6).

These results indicate that the aging of MPs promotes additional interactions [55]. The presence of ions in enriched water intensified ATZ retention, possibly due to the stabilization of chemical interactions between the contaminant and oxygenated functional groups in AMPs.

Isothermal analysis using the Langmuir and Freundlich models indicated that both models were applicable to the ATZ adsorption process ($R^2 > 0.95$), suggesting that the interaction between PMPs and AMPs and ATZ in the aquatic environment is not limited to monolayer adsorption but also includes chemical adsorption mechanisms. For AMPs, the maximum adsorption capacity (Q_m) was significantly higher than in PMPs, demonstrating that aging alters the surface properties of MPs, increasing their affinity for ATZ.

The binding between chemical compounds and polymers is often associated with adsorption and distribution processes [56]. While fully crystalline polymers allow only surface adsorption, semicrystalline polymers, such as polyethylene MPs, enable both surface adsorption and penetration into their amorphous internal regions [57]. As incomplete crystalline polymers, MPs allow ATZ to be adsorbed on the surface and distributed internally. However, due to the slightly alkaline nature of ATZ [18], electrostatic interactions exert limited influence on the adsorption process.

3.5. *Desorption and Phytotoxicity*

During the desorption stage, ATZ molecules were not detected in the desorbed water from MPs with ATZ, indicating that the remaining fraction of ATZ remained strongly adsorbed onto the MPs' surfaces, particularly in AMPs, and was not released under the experimental conditions. These findings suggest that ATZ sorbed onto MPs was resistant to the desorption processes applied. Consequently, phytotoxicity tests were conducted in the presence of MP particles with adsorbed ATZ.

3.5.1. Germination

The germination assays demonstrated significant reductions in the number of germinated *Lactuca sativa* seeds in response to treatments with MPs and ATZ. Treatments combining AMPa and ATZ exhibited the highest reductions, with 34% in DW and 30% in EW. Treatments with PMPs and ATZ showed lower reductions, indicating a less severe impact compared to AMPs. The negative control and treatments with PMPs alone (without ATZ) exhibited germination rates close to the reference values, indicating a lower phytotoxic impact. Figure 7 illustrates the distribution of the mean germination values across different treatments and aquatic matrices.

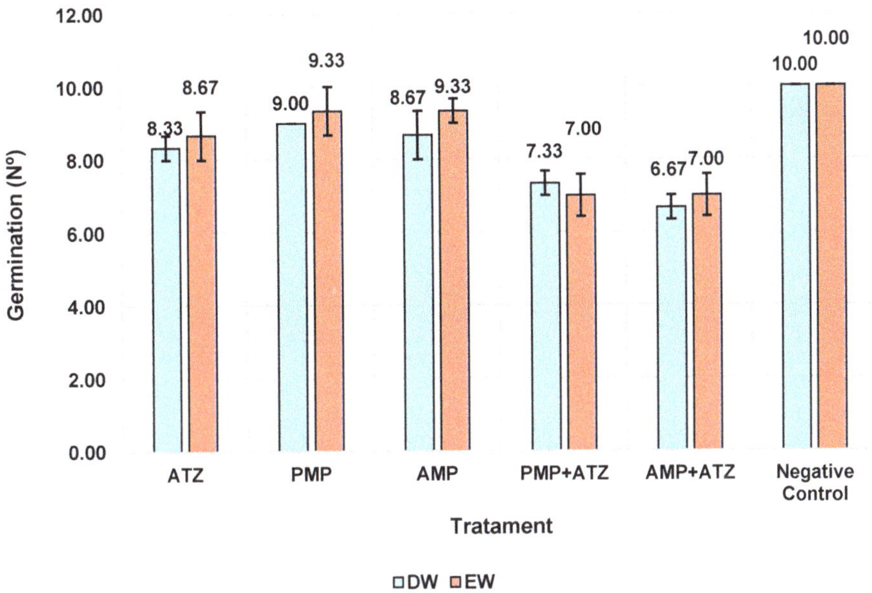

Figure 7. Mean germination numbers with confidence intervals by treatment and medium.

3.5.2. Dry Mass

The dry mass results for *Lactuca sativa* revealed significant variations across different treatments and aquatic matrices. The highest inhibition percentages were observed in treatments with AMPs combined with ATZ (AMP+ATZ), reaching 87.5% in DW and 78.2% in EW (Figure 8). These values were significantly higher than those of other treatments, as indicated by statistical tests ($p < 0.05$).

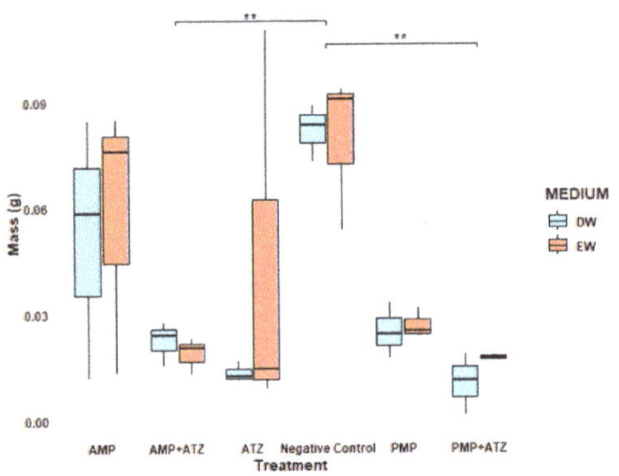

Figure 8. A comparison of *Lactuca sativa* dry mass (g) under different treatments in two aquatic matrices (DW; EW). The treatments include ATZ alone, a negative control, AMPs, AMPs with ATZ (AMP+ATZ), PMPs, and PMPs with ATZ (PMP+ATZ). The bars represent interquartile ranges, while the dots indicate extreme values. The asterisks (*) denote statistically significant differences, ** very significant.

Treatments with PMP+ATZ also exhibited high inhibition, with 84.0% in DW and 72.5% in EW, showing statistically significant differences compared to the treatments with MP or ATZ alone. Treatments with MPs alone showed lower inhibition values: for AMPs,

72.9% in DW and 66.6% in EW, while for PMPs, the inhibition values were 69.6% in DW and 48.5% in EW (Figure 9).

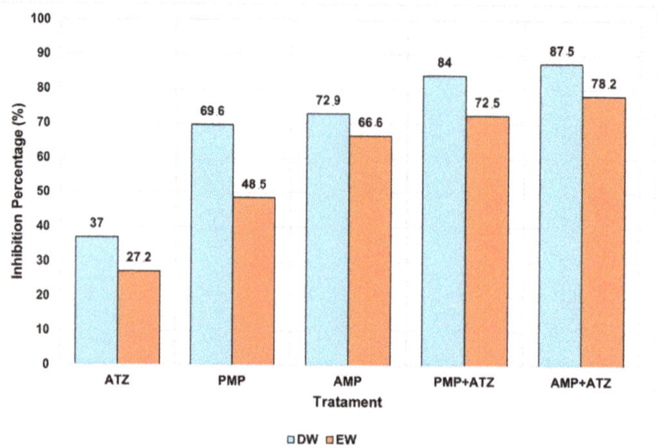

Figure 9. The percentage inhibition of *Lactuca sativa* dry mass under different treatments (ATZ, PMP, AMP, PMP+ATZ, and AMP+ATZ) in DW and EW. The treatments include ATZ alone, AMPs, AMPs with ATZ (AMP+ATZ), PMPs, and PMPs with ATZ (PMP+ATZ).

Conversely, the negative control and ATZ alone presented the lowest inhibition percentages, with 37.0% in DW and 27.2% in EW, showing no statistically significant differences between them. Statistical analysis indicated that the presence of MPs, either alone or combined with ATZ, significantly increased the dry mass inhibition compared to the control, with the AMP+ATZ treatment exhibiting the greatest impact.

3.5.3. Fresh Mass

The fresh mass analysis revealed consistent patterns among the different treatments. Treatments combining AMPs and ATZ exhibited the highest fresh mass inhibition percentages, reaching 58.5% in DW and 46.7% in EW. These values highlight the pronounced impact of these treatments compared to the others (Figure 9). Statistical analysis confirmed that the AMP+ATZ combination was highly significant ($p = 0.00007$) compared to the ATZ-only treatment, demonstrating the synergistic interaction between AMPs and ATZ.

Treatments with PMPs combined with ATZ also showed significant inhibition, with percentages of 39.8% in DW and 38.4% in EW, though these were lower than those observed for AMP+ATZ. These results suggest that PMPs have a lower capacity to amplify the phytotoxic effects of ATZ compared to AMPs. Statistically, the effects of PMP+ATZ were highly significant ($p = 0.00219$) when compared to the ATZ-only treatment (Figure 10).

The isolated AMP and PMP treatments exhibited lower levels of fresh mass inhibition. For AMP, the values were 48.3% in DW and 37.8% in EW, with a statistically significant difference ($p = 0.00224$) compared to the negative control. For PMPs, the inhibition values were 37.2% in DW and 33.5% in EW, also showing high statistical significance ($p = 0.0006$).

Overall, the combined treatments, particularly AMP+ATZ, produced the most pronounced phytotoxic effects, with statistically significant differences compared to the negative controls and isolated treatments. Additionally, adjusted *p*-values reinforced the robustness of the findings, highlighting the role of AMS in amplifying the toxic effects of ATZ (Figure 11).

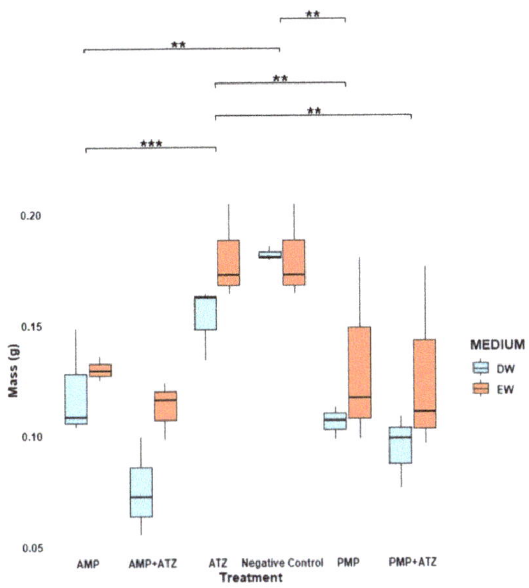

Figure 10. Comparison of the fresh mass (g) of Lactuca sativa under different treatments in two aquatic matrices (DW and EW). Treatments include ATZ alone, a negative control, AMP, AMP +ATZ, PMP, and PMP+ATZ. Bars represent interquartile ranges, and dots indicate extreme values. The asterisks (*) denote statistically significant differences, ** very significant, and *** highly significant.

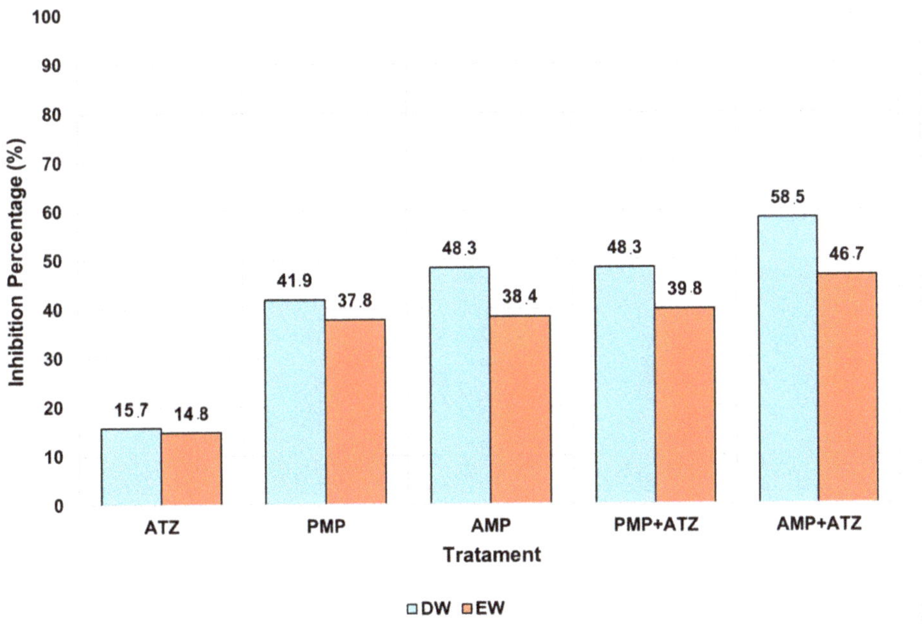

Figure 11. The percentage inhibition of *Lactuca sativa* fresh mass under different treatments (ATZ, PMP, AMP, PMP+ATZ, and AMP+ATZ) in DW and EW.

The results (Figures 7–11) show that ATZ alone exhibited relatively low inhibition of lettuce root growth, whereas ATZ adsorbed onto AMPs significantly intensified the phytotoxic effects. The phytotoxicity assays demonstrated that root elongation and biomass were more severely affected in the AMP+ATZ treatment compared to ATZ alone, suggesting that the presence of AMPs in the liquid medium enhances the bioavailability and toxicity of ATZ.

3.5.4. SEM Analysis of Radicles and Toxicological Classification

Scanning electron microscopy (SEM) analysis and experimental data revealed that, despite the desorption assays failing to detect significant ATZ concentrations, the phytotoxicity assays demonstrated notable differences between treatments with and without the herbicide. This difference was evident even in the absence of detectable ATZ levels in DW and EW, suggesting that while ATZ was adsorbed onto MPs in small amounts, it formed strong bonds and remained retained. However, its release appears to occur only through direct contact with plant radicles.

The images presented in Figure 12 illustrate the interaction between *Lactuca sativa* radicles and MPs, revealing the adhesion of MPs to root surfaces. These images show the distribution of MPs predominantly at the radicle tips, supporting the hypothesis that direct physical contact was the crucial factor for localized ATZ release.

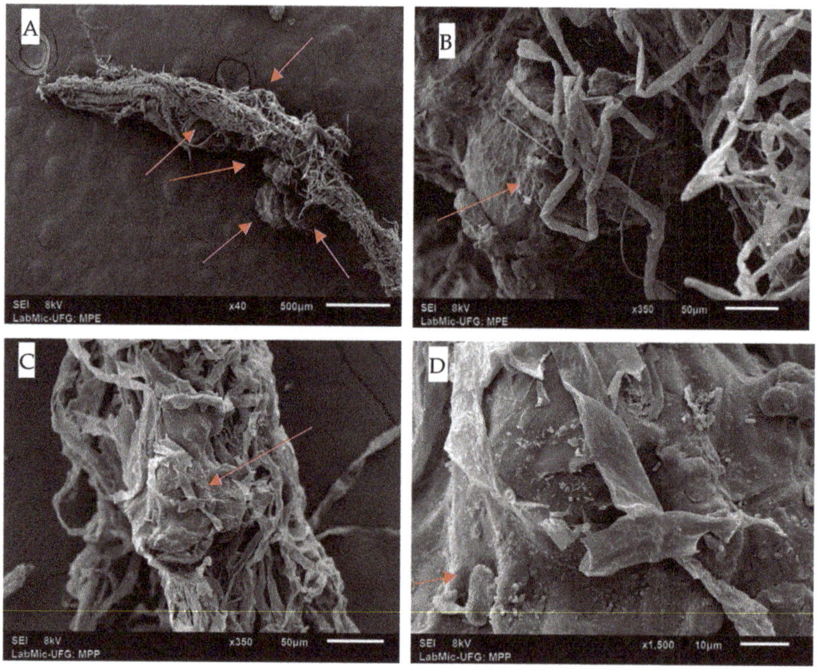

Figure 12. SEM images showing the interaction of MPs with *Lactuca sativa* roots: (**A**) Overview of AMPs adhering to the root surface at 40× magnification (scale bar = 500 µm); (**B**) close-up of radicle entanglement with AMPs, showing adhesion to root structures at 350× magnification (scale bar = 50 µm); (**C**) PMPs interacting with the root surface, showing dispersed and adhered fibers at 350× magnification (scale bar = 50 µm); (**D**) accumulation of PMPs on the root surface, indicating a higher adhesion density at 1500× magnification (scale bar = 10 µm). The red arrows indicate the surface of the microplastics or microplastics involved in the radicles.

Even at residual concentrations, ATZ adsorbed onto MPs was sufficient to cause adverse effects in *Lactuca sativa*. Significant reductions in the germination rates and root length were observed, indicating that small amounts of ATZ retained by AMPs can be biologically relevant, as presented in Table 4. Treatments combining AMPs and ATZ in DW (AMP+ATZ+DW) exhibited the highest inhibition rates, with reductions of 58.52% in fresh mass and 87.50% in dry mass. In enriched water (AMP+ATZ -+ EW), the inhibition values were 46.73% for fresh mass and 78.19% for dry mass (Figures 10 and 11).

Table 4. Root length evolution.

	Treatments	Average	Average Control	% Germinated	Germin. Control (%)	IER	IGN	Toxicity
DW	ATZ	2.49	3.07	83	100	−0.19	−0.17	Low
	PMP	2.79		90		−0.09	−0.10	Low
	AMP	2.66		87		−0.13	−0.13	Low
	PMP+ATZ	1.82		73		−0.41	−0.27	Moderate T
	AMP+ATZ	1.68		67		−0.45	−0.33	Moderate
	Treatments	**Average**	**Average Control**	**% Germinated**	**Germi. Control (%)**	**IER**	**IGN**	**Toxicity**
EW	ATZ	2.72	3.07	87	100	−0.11	−0.13	Low
	PMP	2.75		93		−0.10	−0.07	Low
	AMP	2.65		93		−0.14	−0.07	Low
	PMP+ATZ	1.87		70		−0.39	−0.30	Moderate
	AMP+ATZ	1.86		70		−0.39	−0.30	Moderate

The interaction between MPs with adsorbed ATZ and plant roots can be attributed to multiple physicochemical and biological mechanisms. The adhesion of MPs to root surfaces is likely governed by hydrophobic interactions, root exudate binding, biofilm formation, electrostatic effects, and water-mediated transport. Atrazine is a hydrophobic herbicide (log Kow = 2.61), and when adsorbed onto AMPs, it may alter their surface properties, increasing their hydrophobicity and potentially enhancing root adhesion [58]. Since plant root surfaces contain hydrophobic domains, particularly due to cutin, suberin, and lignin in the cell walls [59], it is possible that hydrophobic attraction played a role in the retention of ATZ-loaded MPs on root surfaces. Figure 13 shows the toxicity zones by treatment and medium

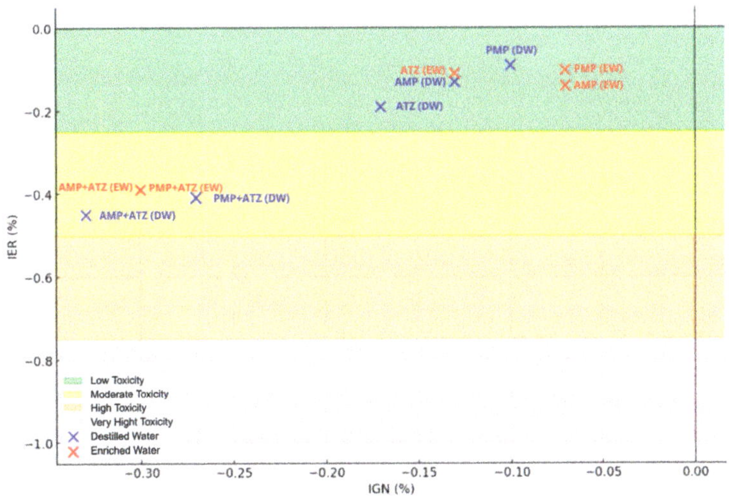

Figure 13. Toxicity zones: analysis by treatment and medium.

Additionally, root exudates, which include proteins, polysaccharides, and phenolic compounds, can mediate MP–root interactions by acting as binding agents [60]. Phenolic exudates may form π-π interactions with ATZ molecules, further enhancing MP attachment to the root surface. Furthermore, mucilage, a gel-like polysaccharide secreted by roots, has been shown to trap particulate matter, including MPs, increasing their retention near the rhizosphere [61].

Another relevant factor is biofilm formation, which can influence MP–root interactions. Rhizosphere microorganisms readily colonize MP surfaces, forming biofilms that enhance adhesion [62]. These biofilms create a sticky extracellular matrix, which may further promote MP retention within root hairs or porous regions of the root surface.

From a size-dependent perspective, MPs in the nanometer to low-micrometer range may enter roots through passive apoplastic pathways (spaces between cells) or active endocytosis, particularly in root hairs [63]. Previous studies have demonstrated that nanoplastics (<100 nm) can be taken up by plant roots, reaching vascular tissues in *Arabidopsis thaliana* [64]. Although the MPs used in this study are in the micrometer range, they may still adhere to or become embedded within root tissues under specific conditions.

The water uptake process may also facilitate MP–root interactions. MPs suspended in water can be transported toward root surfaces via capillary action, particularly under conditions of high transpiration rates. Once near the root zone, MPs may be physically retained within root hairs or adhered to due to localized charge variations influenced by pH gradients in the rhizosphere [65].

These mechanisms collectively suggest that aged and pristine MPs with adsorbed ATZ may interact with plant roots through a combination of hydrophobic effects, biochemical interactions, microbial colonization, and water flow dynamics. Further research is necessary to determine the extent of MP penetration into root tissues and whether these interactions affect a plant's physiology, nutrient uptake, or toxicity responses.

These findings align with the mode of action of ATZ, a herbicide that inhibits photosynthesis by blocking electron transfer in Photosystem II, thereby reducing energy production and the essential organic compounds necessary for plant growth [6]. This disruption is intensified by interactions with MPs, particularly AMPs, which have a greater capacity to retain and release ATZ through direct contact with plant radicles. This dynamic highlights how MPs, both PMPs and AMPs, act as transport vectors, amplifying localized exposure and exacerbating phytotoxic impacts [66].

Phytotoxicity data also revealed significant differences in the germination indices and root lengths of *Lactuca sativa*, underscoring the deleterious effects caused by AMPs associated with ATZ [20,67].

4. Conclusions

This study investigated the sorption behavior of ATZ on PMP and AMP polyethylene MPs, analyzing its environmental and phytotoxic implications. The results demonstrated that AMPs exhibited a higher ATZ adsorption capacity than PMPs, which can be attributed to the structural modifications caused by photodegradation, such as the formation of oxygenated functional groups and increased surface roughness, as confirmed by FTIR and SEM analyses.

The phytotoxicity assays revealed that ATZ adsorbed onto MPs significantly inhibited the germination of *Lactuca sativa*. This effect was more pronounced for the AMPs compared to the PMPs or ATZ alone, indicating that the AMPs act as an enhanced vector for ATZ delivery to plant systems. Experimental observations suggest that the increased phytotoxicity stems from both the stronger physicochemical interactions between AMPs and ATZ, which enhance herbicide retention and transfer, and the direct interference of

MPs in root physiology. Specifically, exposure to AMP-bound ATZ resulted in reduced root elongation and altered root morphology, likely due to oxidative stress and metabolic disruption caused by prolonged ATZ contact.

Furthermore, our findings demonstrate that MPs in the rhizosphere can physically interact with root surfaces through adhesion mechanisms influenced by root exudates, mucilage-mediated attachment, and capillary forces. This interaction may lead to mechanical stress, blockage of root hairs, and disruption of water and nutrient uptake [68]. Additionally, AMP-induced biofilm formation could alter the microbial communities critical for plant health, further compounding the toxic effects. These observations align with previous studies [69,70] suggesting that AMPs exhibit a higher affinity for plant roots due to surface oxidation and charge modifications, intensifying both root entrapment and pollutant uptake.

From a chemical perspective, the adsorption of ATZ onto MPs increases its environmental persistence and bioavailability in the root zone, thereby prolonging plant exposure to this herbicide. This extended exposure was experimentally linked to increased oxidative stress markers in plant tissues and hindered seedling development. These results indicate that MPs not only serve as passive carriers of ATZ but also actively modulate its environmental fate and phytotoxic effects. Given that AMPs adsorb more ATZ than PMPs, their presence in contaminated environments could exacerbate the long-term ecological risks associated with pesticide exposure, particularly in agricultural regions with intensive herbicide application.

Moreover, our results highlight that MPs significantly enhance the persistence of ATZ in aquatic environments, thereby increasing its bioavailability to aquatic organisms. This interaction may amplify the ecological risks posed by both pollutants, particularly in agricultural runoff areas where MPs and ATZ co-occur. Additionally, ATZ was observed to exert direct toxic effects on plants through root contact, raising concerns about its potential uptake in crops and subsequent impacts on plant health and productivity. The presence of MPs further modulates this process by altering ATZ mobility and bioavailability in soil and water matrices, emphasizing the need for further research on its implications for food safety.

Overall, these findings underscore the dual threat posed by MP-bound ATZ, as MP accumulation in the rhizosphere induces mechanical stress and disrupts root nutrient uptake, while ATZ release from MPs enhances phytotoxicity through direct root exposure.

Future research should focus on the mechanisms underlying MP–root interactions, particularly concerning MP penetration into root tissues and its effects on plant metabolism and nutrient absorption. Additionally, assessing the potential entry of MP-bound ATZ into food chains is crucial for understanding its broader ecological and human health implications.

Also, our results indicate that MPs significantly enhance the persistence of ATZ in aquatic environments, which could increase its bioavailability to aquatic organisms. This interaction may exacerbate the long-term ecological risks associated with both pollutants, particularly in agricultural runoff areas. Additionally, ATZ has been revealed to act directly on plants through root contact, which raises concerns about its uptake in crops and potential impacts on plant health and productivity. The presence of MPs may further influence this process by altering the availability and mobility of ATZ in soil and water matrices. Future studies should explore the potential for MP-bound ATZ to enter food chains, as well as mitigation strategies for reducing such contamination

These insights are critical for designing sustainable environmental management strategies aimed at reducing MP pollution and minimizing pesticide-associated risks in terrestrial and aquatic ecosystems.

Author Contributions: Conceptualization, F.J.C.T. and L.G.X.d.S.; methodology, F.J.C.T., A.R.C. and L.G.X.d.S.; validation, F.J.C.T., L.G.X.d.S. and R.M.F.C.; formal analysis F.J.C.T., L.G.X.d.S. and R.M.F.C.; investigation, F.J.C.T., L.G.X.d.S., K.C.d.S. and A.R.C.; resources, F.J.C.T.; data curation, F.J.C.T. and L.G.X.d.S.; writing—original draft preparation, F.J.C.T. and L.G.X.d.S.; writing—review and editing, F.J.C.T., L.G.X.d.S. and R.M.F.C.; visualization, L.G.X.d.S.; supervision, F.J.C.T.; project administration F.J.C.T.; funding acquisition, F.J.C.T. All authors have read and agreed to the published version of the manuscript.

Funding: This research received no external funding, and the APC was granted by the post-graduation program on Sanitary and Environmental Engineering of the Federal University of Goiás.

Institutional Review Board Statement: Not applicable.

Informed Consent Statement: Not applicable.

Data Availability Statement: Data will be available upon request to the authors.

Conflicts of Interest: The authors declare no conflicts of interest.

Abbreviations

The following abbreviations are used in this manuscript:

ATZ	Atrazine
MP	Microplastic
AMP	Aged microplastic
PMP	Pristine microplastic
DW	Distilled water
EW	Enriched water
CI	Carbonyl index
qt	Adsorption capacity
Co	Initial concentration
Ct	Final concentration
m	Mass of microplastic
PFO	Pseudo-first-order
PSO	Pseudo-second-order
ICR	Root Growth Index.
CRA	Root Growth in Sample.
CRC	Root Growth in the Control.
IG	Germination Index.
SGA	Number of Germinated Seeds in Sample.
SGC	Number of Germinated Seeds in Negative Control.
IER	Normalized Root Growth Percentage Index.
CMRCA	Average Radicle Length in Each Sample.
CMRCN	Average Radicle Length of Germinated Seeds in the Control.
IGN	Normalized Residual Germination Percentage Index
GERM	Germinated Seeds.
SEM	Scanning electron microscopy
FTIR	Fourier transform infrared spectroscopy
PE	Polystyrene
ZP	Zeta potential

References

1. Leusch, F.D.; Lu, H.-C.; Perera, K.; Neale, P.A.; Ziajahromi, S. Analysis of the literature shows a remarkably consistent relationship between size and abundance of microplastics across different environmental matrices. *Environ. Pollut.* **2023**, *319*, 120984. [CrossRef] [PubMed]
2. WebWire. Center for Sustainability of Credit Suisse Publishes Study on Plastic Pollution. 2023. Available online: https://www.webwire.com/ViewPressRel.asp?aId=305845 (accessed on 11 February 2025).

3. Lara, L.Z.; Bertoldi, C.; Alves, N.M.; Fernandes, A.N. Sorption of endocrine disrupting compounds onto polyamide microplastics under different environmental conditions: Behaviour and mechanism. *Sci. Total Environ.* **2021**, *796*, 148983. [CrossRef] [PubMed]
4. Jiang, X.; Chen, H.; Liao, Y.; Ye, Z.; Li, M.; Klobučar, G. Ecotoxicity and genotoxicity of polystyrene microplastics on higher plant *Vicia faba*. *Environ. Pollut.* **2019**, *250*, 831–838. [CrossRef] [PubMed]
5. Patidar, K.; Alluhayb, A.H.; Younis, A.M.; Dumka, U.; Ambade, B. Investigation of microplastic contamination in the gastrointestinal tract of fish: A comparative study of various freshwater species. *Phys. Chem. Earth Parts A/B/C* **2024**, *136*, 103760. [CrossRef]
6. Narayanan, M.; Devarayan, K.; Verma, M.; Selvaraj, M.; Ghramh, H.A.; Kandasamy, S. Assessing the ecological impact of pesticides/herbicides on algal communities: A comprehensive review. *Aquat. Toxicol.* **2024**, *268*, 106851. [CrossRef]
7. Wieland, S.; Balmes, A.; Bender, J.; Kitzinger, J.; Meyer, F.; Ramsperger, A.F.; Roeder, F.; Tengelmann, C.; Wimmer, B.H.; Laforsch, C.; et al. From properties to toxicity: Comparing microplastics to other airborne microparticles. *J. Hazard. Mater.* **2022**, *428*, 128151. [CrossRef]
8. Nguyen, M.-K.; Rakib, R.J.; Lin, C.; Hung, N.T.Q.; Le, V.-G.; Nguyen, H.-L.; Malafaia, G.; Idris, A.M. A comprehensive review on ecological effects of microplastic pollution: An interaction with pollutants in the ecosystems and future perspectives. *TrAC Trends Anal. Chem.* **2023**, *168*, 117294. [CrossRef]
9. Rai, P.K.; Sonne, C.; Brown, R.J.C.; Younis, S.A.; Kim, K.-H. Adsorption of environmental contaminants on micro- and nano-scale plastic polymers and the influence of weathering processes on their adsorptive attributes. *J. Hazard. Mater.* **2022**, *427*, 127903. [CrossRef]
10. Menéndez-Pedriza, A.; Jaumot, J. Interaction of Environmental Pollutants with Microplastics: A Critical Review of Sorption Factors, Bioaccumulation and Ecotoxicological Effects. *Toxics* **2020**, *8*, 40. [CrossRef]
11. Khalid, N.; Aqeel, M.; Noman, A.; Khan, S.M.; Akhter, N. Interactions and effects of microplastics with heavy metals in aquatic and terrestrial environments. *Environ. Pollut.* **2021**, *290*, 118104. [CrossRef]
12. Verla, A.W.; Enyoh, C.E.; Verla, E.N.; Nwarnorh, K.O. Microplastic–toxic chemical interaction: A review study on quantified levels, mechanism and implication. *SN Appl. Sci.* **2019**, *1*, 1400. [CrossRef]
13. Beriot, N.; Zomer, P.; Zornoza, R.; Geissen, V. A laboratory comparison of the interactions between three plastic mulch types and 38 active substances found in pesticides. *PeerJ* **2020**, *8*, e9876. [CrossRef] [PubMed]
14. Lv, Z.-Y.; Wang, C.-Y.; Liu, S.-X.; Xu, Y.-C.; Xu, K. Effect of microplastic coexistence conditions on the environmental behavior of atrazine on soil. *Appl. Ecol. Env. Res.* **2023**, *21*, 3659–3673. [CrossRef]
15. Tang, K.H.D. Effects of Microplastics on Bioavailability, Persistence and Toxicity of Plant Pesticides: An Agricultural Perspective. *Agriculture* **2025**, *15*, 356. [CrossRef]
16. Ghimire, N.; Woodward, R.T. Under- and over-use of pesticides: An international analysis. *Ecol. Econ.* **2013**, *89*, 73–81. [CrossRef]
17. Zhou, W.; Li, M.; Achal, V. A comprehensive review on environmental and human health impacts of chemical pesticide usage. *Emerg. Contam.* **2025**, *11*, 100410. [CrossRef]
18. Guo, J.; Du, Y.; Yang, L.; Luo, Y.; Zhong, G.; Zhao, H.-M.; Liu, J. Effects of microplastics on the environmental behaviors of the herbicide atrazine in soil: Dissipation, adsorption, and bioconcentration. *J. Hazard. Mater.* **2024**, *465*, 133085. [CrossRef]
19. Wang, T.; Zhou, Y.; Xue, Y.; Sang, T.; Ren, L.; Chen, S.; Liu, J.; Mei, M.; Li, J. Pyrolysis of Hydrothermally Dewatering Sewage Sludge: Highly Efficient Peroxydisulfate Activation of Derived Biochar to Degrade Diclofenac. *Environ. Pollut.* **2022**, *313*, 120176. [CrossRef]
20. Sun, S.; Sui, H.; Xu, L.; Zhang, J.; Wang, D.; Zhou, Z. Effect of freeze-thaw cycle aging and high-temperature oxidation aging on the sorption of atrazine by microplastics. *Environ. Pollut.* **2022**, *307*, 119434. [CrossRef]
21. Bombardi, L.M. *Agrotóxicos e Colonialismo Químico*; Editora Elefante: Minas Gerais, Brazil, 2023.
22. WHO. *Occupational Exposures in Insecticide Application, and Some Pesticides*; International Agency for Research on Cancer: Lyon, France, 1991.
23. Ghirardelli, A.; Otto, S.; Masin, R.; Bano, C.; Altissimo, L.; Russo, S.; Zanin, G. Thirty-year monitoring of s-triazine herbicide contamination in the aquifer north of Vicenza (north-east Italy). *Sci. Total Environ.* **2021**, *752*, 141647. [CrossRef]
24. Zeng, Z.; Jia, B.; Liu, X.; Chen, L.; Zhang, P.; Qing, T.; Feng, B. Adsorption behavior of triazine pesticides on polystyrene microplastics aging with different processes in natural environment. *Environ. Pollut.* **2024**, *356*, 124319. [CrossRef] [PubMed]
25. BiologyInsights. Atrazine's Effects on Aquatic Ecosystems and Organisms. 2025. Available online: https://biologyinsights.com/atrazines-effects-on-aquatic-ecosystems-and-organisms/ (accessed on 7 March 2025).
26. Fromang, S. Effecs of atrazine on non-targed organisms. Technical Paper. University of Florida. 2024. Available online: https://soils.ifas.ufl.edu/ (accessed on 5 March 2025).
27. Conley, M. Atrazine, an Endocrine-Disrupting Herbicide Banned in Europe, Is Widely Used in the U.S. *U.S. Right to Know*. 7 January 2025. Available online: https://usrtk.org/pesticides/atrazine/ (accessed on 5 March 2025).
28. Li, H.; Wang, F.; Li, J.; Deng, S.; Zhang, S. Adsorption of three pesticides on polyethylene microplastics in aqueous solutions: Kinetics, isotherms, thermodynamics, and molecular dynamics simulation. *Chemosphere* **2021**, *264*, 128556. [CrossRef] [PubMed]

29. Sun, S.; Yang, X.; Xu, L.; Zhang, J.; Wang, Y.; Zhou, Z. Atrazine sorption on biodegradable microplastics: Significance of microbial aging. *Sci. Total Environ.* **2023**, *862*, 160904. [CrossRef]
30. Dias, M.A.; Batista, P.R.; Ducati, L.C.; Montagner, C.C. Insights into sorption and molecular transport of atrazine, testosterone, and progesterone onto polyamide microplastics in different aquatic matrices. *Chemosphere* **2023**, *318*, 137949. [CrossRef]
31. de Albuquerque, F.P.; de Oliveira, J.L.; Moschini-Carlos, V.; Fraceto, L.F. An overview of the potential impacts of atrazine in aquatic environments: Perspectives for tailored solutions based on nanotechnology. *Sci. Total Environ.* **2020**, *700*, 134868. [CrossRef]
32. Graymore, M.; Stagnitti, F.; Allinson, G. Impacts of atrazine in aquatic ecosystems. *Environ. Int.* **2001**, *26*, 483–495. [CrossRef]
33. Singh, S.; Kumar, V.; Chauhan, A.; Datta, S.; Wani, A.B.; Singh, N.; Singh, J. Toxicity, degradation and analysis of the herbicide atrazine. *Environ. Chem. Lett.* **2018**, *16*, 211–237. [CrossRef]
34. Ainali, N.M.; Bikiaris, D.N.; Lambropoulou, D.A. Aging effects on low- and high-density polyethylene, polypropylene and polystyrene under UV irradiation: An insight into decomposition mechanism by Py-GC/MS for microplastic analysis. *J. Anal. Appl. Pyrolysis* **2021**, *158*, 105207. [CrossRef]
35. Albergamo, V.; Wohlleben, W.; Plata, D.L. Photochemical weathering of polyurethane microplastics produced complex and dynamic mixtures of dissolved organic chemicals. *Environ. Sci. Process. Impacts* **2023**, *25*, 432–444. [CrossRef]
36. Mañas, P.; De las Heras, J. Phytotoxicity test applied to sewage sludge using *Lactuca sativa* L. and *Lepidium sativum* L. seeds. *Int. J. Environ. Sci. Technol.* **2018**, *15*, 273–280. [CrossRef]
37. Guari, E.B.; de Almeida, É.J.R.; Martiarena, M.d.J.S.; Yamagami, N.S.; Corso, C.R. Azo Dye Acid Blue 29: Biosorption and Phytotoxicity Test. *Water, Air Soil Pollut.* **2015**, *226*, 361. [CrossRef]
38. Orlina, E.R.; Besana, N.A.D.; Señeris, G.T.; Capiña, K.C. Phytotoxicity Assessment of Biofertilizer Produced from Bioreactor Composting Technology Using Lettuce (*Lactuca sativa* L.) Seeds. *OJE* **2023**, *13*, 257–270. [CrossRef]
39. Bagur-González, M.G.; Estepa-Molina, C.; Martín-Peinado, F.; Morales-Ruano, S. Toxicity assessment using *Lactuca sativa* L. bioassay of the metal(loid)s As, Cu, Mn, Pb and Zn in soluble-in-water saturated soil extracts from an abandoned mining site. *J. Soils Sediments* **2011**, *11*, 281–289. [CrossRef]
40. Stropan, K.V.S.; de Souza, E.C.; Viana, L.O. Fitotoxicidade de efluente da indústria cervejeira em sementes de Lactuca sativa L. | Revista Internacional de Ciências. 2025. Available online: https://www.e-publicacoes.uerj.br/index.php/ric/article/view/30072 (accessed on 10 February 2025).
41. Song, Y.K.; Hong, S.H.; Jang, M.; Han, G.M.; Jung, S.W.; Shim, W.J. Combined Effects of UV Exposure Duration and Mechanical Abrasion on Microplastic Fragmentation by Polymer Type. *Environ. Sci. Technol.* **2017**, *51*, 4368–4376. [CrossRef]
42. Kalčíková, G.; Skalar, T.; Marolt, G.; Jemec Kokalj, A. An environmental concentration of aged microplastics with adsorbed silver significantly affects aquatic organisms. *Water Res.* **2020**, *175*, 115644. [CrossRef]
43. Chaudhary, A.K.; Vijayakumar, R.P. Effect of chemical treatment on biological degradation of high-density polyethylene (HDPE). *Environ. Dev. Sustain.* **2020**, *22*, 1093–1104. [CrossRef]
44. Galant, C.; Fayolle, B.; Kuntz, M.; Verdu, J. Thermal and radio-oxidation of epoxy coatings. *Prog. Org. Coat.* **2010**, *69*, 322–329. [CrossRef]
45. Celina, M.C.; Linde, E.; Martinez, E. Carbonyl Identification and Quantification Uncertainties for Oxidative Polymer Degradation. *Polym. Degrad. Stab.* **2021**, *188*, 109550. [CrossRef]
46. Zidar, P.; Kühnel, D.; Škapin, A.S.; Skalar, T.; Drobne, D.; Škrlep, L.; Mušič, B.; Kokalj, A.J. Comparing the effects of pristine and UV–VIS aged microplastics: Behavioural response of model terrestrial and freshwater crustaceans. *Ecotoxicol. Environ. Saf.* **2024**, *285*, 117020. [CrossRef]
47. Bhagat, K.; Barrios, A.C.; Rajwade, K.; Kumar, A.; Oswald, J.; Apul, O.; Perreault, F. Aging of microplastics increases their adsorption affinity towards organic contaminants. *Chemosphere* **2022**, *298*, 134238. [CrossRef]
48. Nascimento, R.F.; Lima, A.C.; Vidal, C.B.; Melo, D.D.; Raulino, G.S. *Adsorção: Aspectos teóricos e aplicações ambientais*; Imprensa Universitária: Fortaleza, Brazil, 2020.
49. Dolan, T.; Howsam, P.; Parsons, D.; Whelan, M. Is the EU Drinking Water Directive Standard for Pesticides in Drinking Water Consistent with the Precautionary Principle? *Environ. Sci. Technol.* **2013**, *47*, 4999–5006. [CrossRef] [PubMed]
50. Kalra, A.; Tugcu, N.; Cramer, S.M.; Garde, S. Salting-In and Salting-Out of Hydrophobic Solutes in Aqueous Salt Solutions. *J. Phys. Chem. B* **2001**, *105*, 6380–6386. [CrossRef]
51. Xia, Y.; Zhou, J.-J.; Gong, Y.-Y.; Li, Z.-J.; Zeng, E.Y. Strong influence of surfactants on virgin hydrophobic microplastics adsorbing ionic organic pollutants. *Environ. Pollut.* **2020**, *265*, 115061. [CrossRef] [PubMed]
52. Wang, Y.; Liu, C.; Wang, F.; Sun, Q. Behavior and mechanism of atrazine adsorption on pristine and aged microplastics in the aquatic environment: Kinetic and thermodynamic studies. *Chemosphere* **2022**, *292*, 133425. [CrossRef]
53. Chen, C.; Wei, F.; Ye, L.; Wang, Y.; Long, L.; Xu, C.; Xiao, Y.; Wu, J.; Xu, M.; He, J.; et al. Adsorption of Cu2+ by UV aged polystyrene in aqueous solution. *Ecotoxicol. Environ. Saf.* **2022**, *232*, 113292. [CrossRef]

54. Zhang, M.; Liu, N.; Hou, L.; Li, C.; Li, C. Adsorption behaviors of chlorpyrifos on UV aged microplastics. *Mar. Pollut. Bull.* **2023**, *190*, 114852. [CrossRef]
55. Zhu, F.; Zhu, C.; Wang, C.; Gu, C. Occurrence and Ecological Impacts of Microplastics in Soil Systems: A Review. *Bull. Environ. Contam. Toxicol.* **2019**, *102*, 741–749. [CrossRef]
56. Xing, B.; Pignatello, J.J. Dual-Mode Sorption of Low-Polarity Compounds in Glassy Poly(Vinyl Chloride) and Soil Organic Matter. *Environ. Sci. Technol.* **1997**, *31*, 792–799. [CrossRef]
57. Haider, N.; Karlsson, S. Loss of Chimassorb 944 from LDPE and identification of additive degradation products after exposure to water, air and compost. *Polym. Degrad. Stab.* **2001**, *74*, 103–112. [CrossRef]
58. Wang, F.; Wong, C.S.; Chen, D.; Lu, X.; Wang, F.; Zeng, E.Y. Interaction of toxic chemicals with microplastics: A critical review. *Water Res.* **2018**, *139*, 208–219. [CrossRef]
59. Schreiber, L.; Franke, R.; Hartmann, K. Wax and suberin development of native and wound periderm of potato (*Solanum tuberosum* L.) and its relation to peridermal transpiration. *Planta* **2005**, *220*, 520–530. [CrossRef] [PubMed]
60. Jones, D.L.; Nguyen, C.; Finlay, R.D. Carbon flow in the rhizosphere: Carbon trading at the soil–root interface. *Plant Soil* **2009**, *321*, 5–33. [CrossRef]
61. de Souza Machado, A.A.; Lau, C.W.; Till, J.; Kloas, W.; Lehmann, A.; Becker, R.; Rillig, M.C. Impacts of Microplastics on the Soil Biophysical Environment. *Environ. Sci. Technol.* **2018**, *52*, 9656–9665. [CrossRef]
62. Jin, K.; White, P.J.; Whalley, W.R.; Shen, J.; Shi, L. Shaping an Optimal Soil by Root–Soil Interaction. *Trends Plant Sci.* **2017**, *22*, 823–829. [CrossRef]
63. Li, L.; Luo, Y.; Li, R.; Zhou, Q.; Peijnenburg, W.J.; Yin, N.; Yang, J.; Tu, C.; Zhang, Y. Effective Uptake of Submicrometre Plastics by Crop Plants via a Crack-Entry Mode. *Nat. Sustain.* **2020**, *3*, 929–937. [CrossRef]
64. Azeem, I.; Adeel, M.; Ahmad, M.A.; Shakoor, N.; Jiangcuo, G.D.; Azeem, K.; Ishfaq, M.; Shakoor, A.; Ayaz, M.; Xu, M.; et al. Uptake and Accumulation of Nano/Microplastics in Plants: A Critical Review. *Nanomaterials* **2021**, *11*, 2935. [CrossRef]
65. Bais, H.P.; Weir, T.L.; Perry, L.G.; Gilroy, S.; Vivanco, J.M. The role of root exudates in rhizosphere interactions with plants and other organisms. *Annu. Rev. Plant Biol.* **2006**, *57*, 233–266. [CrossRef]
66. Gan, Q.; Cui, J.; Jin, B. Environmental microplastics: Classification, sources, fates, and effects on plants. *Chemosphere* **2023**, *313*, 137559. [CrossRef]
67. Zhang, K.; Hamidian, A.H.; Tubić, A.; Zhang, Y.; Fang, J.K.; Wu, C.; Lam, P.K. Understanding plastic degradation and microplastic formation in the environment: A review. *Environ. Pollut.* **2021**, *274*, 116554. [CrossRef]
68. Adamczyk, S.; Zantis, L.J.; van Loon, S.; van Gestel, C.A.; Bosker, T.; Hurley, R.; Nizzetto, L.; Adamczyk, B.; Velmala, S. Biodegradable microplastics induce profound changes in lettuce (*Lactuca sativa*) defense mechanisms and to some extent deteriorate growth traits. *Environ. Pollut.* **2024**, *363*, 125307. [CrossRef]
69. Enstone, D.E.; Peterson, C.A.; Ma, F. Root Endodermis and Exodermis: Structure, Function, and Responses to the Environment. *J. Plant Growth Regul.* **2002**, *21*, 335–351. [CrossRef]
70. Jiang, Y.; Hou, L.; Ding, F.; Whalen, J.K. Root mucilage: Chemistry and functions in soil. In *Encyclopedia of Soils in the Environment*, 2nd ed.; Goss, M.J., Oliver, M., Eds.; Academic Press: Oxford, UK, 2023; pp. 332–342.

toxics

Article

Interpretable Machine Learning Models and Symbolic Regressions Reveal Transfer of Per- and Polyfluoroalkyl Substances (PFASs) in Plants: A New Small-Data Machine Learning Method to Augment Data and Obtain Predictive Equations

Yuan Zhang [1,2], Yanting Li [1], Yang Li [1], Lin Zhao [1] and Yongkui Yang [1,*]

1 School of Environmental Science and Engineering, Tianjin University, Tianjin 300350, China; yz685604@gmail.com (Y.Z.); li_yanting666@163.com (Y.L.); liy0128@163.com (Y.L.); zhaolin@tju.edu.cn (L.Z.)
2 Georgia Tech Shenzhen Institute, Tianjin University, Shenzhen 518071, China
* Correspondence: ykyang@tju.edu.cn; Tel.: +86-182-0266-1713

Abstract

Machine learning (ML) techniques are becoming increasingly valuable for modeling the transport of pollutants in plant systems. However, two challenges (small sample sizes and a lack of quantitative calculation functions) remain when using ML to predict migration in hydroponic systems. For the bioaccumulation of per- and polyfluoroalkyl substances, we studied the key factors and quantitative calculation equations based on data augmentation, ML, and symbolic regression. First, feature expansion was performed on the input data after data preprocessing; the most important step was data augmentation. The original training set was expanded nine times by combining the synthetic minority oversampling technique and a variational autoencoder. Subsequently, the four ML models were applied to the test set to predict the selected output parameters. Categorical boosting (CatBoost) had the highest prediction accuracy (R^2 = 0.83). The Shapley Additive Explanation values indicated that molecular weight and exposure time were the most important parameters. We applied three symbolic regression models to obtain accurate prediction equations based on the original and augmented data. Based on augmented data, the high-dimensional sparse interaction equation exhibited the highest accuracy (R^2 = 0.776). Our results indicate that this method could provide crucial insights into absorption and accumulation in plant roots.

Keywords: machine learning; data augmentation; symbolic regression; PFAS bioaccumulation; quantitative prediction

1. Introduction

Robust predictive modeling is imperative for advancing risk assessments, informing regulatory framework development, and formulating effective and sustainable mitigation strategies for emerging contaminants [1,2]. Comprehensive research on the uptake and internal transport of contaminants is crucial for accurately delineating the potential risks to both ecosystems and human health [3]. Root concentration factor (RCF) modeling is important for elucidating the complex dynamics of plant–contaminant interactions [4]. The RCF characterizes the accumulation of a contaminant in the roots of a plant in relation to its concentration in an exposure medium [5]. Among the emerging contaminants, perfluoroalkyl substances (PFASs) pose unique challenges. Compared to 4700 compounds with a wide

 https://doi.org/10.3390/toxics13070579

range of molecular sizes, structures, and functional groups [6], PFASs exhibit extensive heterogeneity, complicating our understanding of their environmental behavior [7–9]. This diversity underscores the importance of elucidating the mechanisms by which PFASs are absorbed and accumulated in plant roots.

Since the 1970s, numerous modeling studies have attempted to correlate the physicochemical properties of contaminants with their uptake by plants. Early approaches relied primarily on the octanol/water partition coefficient as the single predictive parameter. However, this single-parameter strategy is inadequate for PFASs, resulting in low accuracy and limited applicability across different plant species and PFAS variants [10–13]. Subsequently, more advanced compartmental models have been developed to incorporate a wider range of physicochemical and environmental factors to capture the complexity of the contaminant uptake and translocation processes. Despite these advances, these models still face challenges in accurately predicting PFAS behavior, largely because of the unique properties of PFASs, such as their ionic form and variable environmental interactions [14,15].

Machine learning (ML) has experienced a substantial surge in popularity in environmental research due to its efficacy in addressing multivariate problems [16,17]. For example, an RCF database with 246 data points was built for 57 organic compounds in 11 crops, using 15 chemical, soil, and plant features to develop four ML models, i.e., a fully connected neural network (FCNN), gradient-boosted regression trees (GBRT), random forest (RF), and support vector regression (SVR) [18]. FCNN and GBRT performed best, with R^2 values of 0.79 and 0.76 and mean absolute error values of 0.22 and 0.23, respectively.

The prediction of transport factors faces two major challenges. First, "small data," which can be defined as insufficient sample sizes or sample-to-feature ratios below the recommended thresholds [19], directly constrain the predictive ability of ML models in characterizing the transport, dispersion, and distribution of chemicals in the environment. Second, although ML models can achieve highly certain predictive performance, they inherently face interpretability challenges. While post-interpretation methods, such as Shapley Additive Explanation (SHAP) values, have made progress in revealing feature importance and contributions, these techniques provide only approximate explanations of model behavior rather than precise mathematical expressions. From a strictly scientific paradigm, the lack of explicit mathematical expressions makes validating and generalizing the relationship between the model and phenomenon through a theoretical framework challenging. This limits our in-depth understanding of the underlying mechanisms to some extent.

We conducted data augmentation of PFAS transport in the roots of hydroponic plants by combining the synthetic minority oversampling technique (SMOTE) and a variational autoencoder (VAE) model to address the challenges of small data and the low transparency of PFAS translocation studies in hydroponics and solve these two bottlenecks. Three symbolic regression methods were adopted to determine the mathematical equations with the highest predictive accuracy. The specific objectives of this study were to (1) develop a specialized SDML workflow tailored to target values, especially in terms of data augmentation, (2) find the best ML model to predict PFAS translocation in hydroponics, (3) quantify the relative contributions of key drivers affecting the translocation of PFASs in hydroponics, and (4) establish mathematical equations to estimate the target values for different PFASs and plant species from selected quantifiable properties. Therefore, this study presents a new small-data ML method to augment data and obtain predictive equations. RF was used in this study. In addition to using three mainstream integrated models (categorical boosting (CatBoost), the light gradient boosting machine (LightGBM) and extreme gradient boosting (XGBoost)) were used to explore the key drivers for prediction using SHAP analysis. Finally, symbolic regression was implemented to enhance the availability of predictive models using mathematical equations.

2. Materials and Methods

2.1. Data Preprocessing, Derived Feature Construction, and Selection for log RCF

2.1.1. Data Description and Preprocessing

A comprehensive experimental dataset including 616 data points, representing different plants with diverse RCFs, was compiled to predict the plant uptake of PFASs from hydroponics using experimental data from peer-reviewed publications using the Web of Science and Google Scholar from 2009 to 2024, as shown in Figure S1 and Table S1. The features included in this dataset were used as the base features. An iterative imputation algorithm based on an RF regressor [20] was used to fit and impute the data and address the issue of missing data for the target variables in the dataset. Subsequently, the quartiles and interquartile range of the data were calculated, and the range of abnormal values was determined based on the 1.5 times interquartile range (1.5 IQR) rule. A logarithmic transformation was implemented to enhance the symmetry of the data and their propensity to approach a normal distribution, where the trimmed data demonstrated a substantial positive bias (skewness > 1). The detailed construction process is presented in TEXTs S1 and S2, and the encoded variables are listed in Table S2. Finally, the two data sources were integrated to construct a more comprehensive molecular feature space. The chemical identifiers (CIDs) and published physicochemical properties of the compounds were obtained using the PubChem application programming interface (API) [21]. We used the open-source RDKit Python (https://www.rdkit.org/docs/index.html, accessed on 6 September 2024) API and the Ray framework [22] to enable the parallel computation and extraction of a variety of structural features. Finally, the entire dataset was divided into training (75%) and testing (25%) sets.

2.1.2. Derived Feature Construction and Selection

Additional meaningful features must be generated to enhance the model's performance. Moreover, the predictive capability of the original data is augmented by constructing the derived features through a series of nonlinear transformations and interactive constructions [23]. The detailed construction process is presented in TEXT S3.

Feature selection is a critical step in ML and data mining [24,25]. In this study, classical statistics (TEXT S4) were leveraged (including the F-statistic for linear relationships [26], mutual information for nonlinear dependencies [27], and distance correlation for generalized statistical dependencies [28]) to generate the initial indicators of feature relevance. These measures were then normalized, weighted, and combined to yield a comprehensive feature-importance score. This method further integrates stability analysis through bootstrap sampling to determine the consistency of feature importance [29] across different subsets and solves the problem of multicollinearity by penalizing redundant features by using a variance inflation factor (VIF) [30]. Furthermore, the introduction of the maximum information coefficient (MIC) facilitates the detection of complex nonlinear relationships [31], whereas the ReliefF algorithm enhances the discriminative power of features by considering local instance-based assessments [32]. The pseudocode for the design of the entire algorithm is presented in TEXT S5.

2.2. Data Augmentation for log RCF Using Stratified Variational Regression

We combined three data augmentation methods: hierarchical and box-splitting strategies, VAE generation, and SMOTE interpolation. We maintained the original distribution characteristics and effectively solved the sample imbalance problem in each interval by dynamically evaluating the sample distribution of each box and customizing the augmentation ratio. This approach is more robust and adaptive. First, it uses adaptive binning technology [33] to divide the target space into multiple statistical regions and automatically selects a binning strategy based on data skewness. We ensured that each range contained a similar number of samples for unevenly distributed data to maintain statistical balance. For more evenly distributed data, ranges of equal size were created to preserve the physical meaning of the RCF intervals, making the results more interpretable from an environmental perspective. The number of bins was precisely determined through a trade-off between the data scale and bin granularity.

$$n_{bins} = max(6, min(12, n_{train}/12)) \quad (1)$$

Each box captured the local statistical characteristics of the target variable, including the number of samples, mean, standard deviation, and value range. We also evaluated the balance of the sample distribution within each range to determine how much additional data generation was needed.

The algorithm used a dual-pipeline generation strategy to create new samples. SMOTE was the first pipeline used [34]. The synthetic minority oversampling technique for regression (SMOTER) was used to identify similar samples in the feature space and perform intelligent interpolation among them, with particular consideration given to the continuous distribution characteristics of the log RCF. This process creates new chemical property combinations by blending the characteristics from similar existing compounds while adding small variations to increase diversity. This mimicked the natural variation in chemical properties within structural families, which can be formally represented as

$$x_{new} = x_i + \mu(x_{nn} - x_i) + \delta,\ \mu \sim U(0,1) \quad (2)$$

Among them, δ is a slight noise related to the features, which is added proportionally to maintain the scale relationship between the features. The second pipeline utilized a VAE [35] to learn the underlying patterns in the chemical data and generate new samples that follow these patterns but represent novel chemical combinations. This method captures complex nonlinear relationships between molecular properties that simple interpolation may miss. This generative approach ensures that new samples maintain realistic relationships between different molecular descriptors, while expanding the chemical space coverage.

The data generation process was iterative and self-monitored. After each round of sample generation, we evaluated whether the new data maintained realistic chemical relationships and provided a better representation across different RCF ranges. This ensured that our enhanced dataset had more samples and preserved the fundamental chemical and biological relationships that govern the plant uptake of PFASs. The pseudocode for the entire algorithm design is presented in TEXT S6.

2.3. *Development of ML Model for log RCF*

2.3.1. ML Models

We implemented four types of algorithms that are considered the most effective for improving the prediction accuracy in this study [36]. The CatBoost regressor [37], a sophisticated gradient-boosting implementation, has demonstrated superior performance in predicting PFAS bioaccumulation in groundwater [34] through its innovative ordered boosting algorithm, which minimizes prediction shifts arising from target leakage. CatBoost can be computationally intensive for large datasets and may require extensive hyperparameter tuning to achieve optimal performance. The LightGBM [38] uses a histogram-based decision-tree growth strategy that discretizes continuous features into optimal bins, dramatically reducing memory requirements while preserving statistical fidelity. However, LightGBM is sensitive to small datasets and may suffer from overfitting when the number of samples is limited relative to the features. The XGBoost algorithm [39] addresses high-dimensional sparsity and multicollinearity in the molecular descriptor space through its distinctive regularization framework, which combines L1 and L2 penalties. Despite its strengths, XGBoost is prone to overfitting noisy data and requires careful parameter tuning to balance the bias and variance trade-offs. The RF model [40] builds a low-correlation decision tree integration by combining bootstrap aggregation with random feature subset selection at each node split, thereby providing robustness against overfitting and inherent variance reduction through ensemble averaging. However, RF may struggle with extrapolation beyond the training data range and may be biased toward features with more levels of categorical variables. The detailed differences are listed in Table S3.

2.3.2. Hyperparameter Search Using Bayesian Search

The hyperparametric optimization methodology used in this study utilizes a multilevel model-tuning strategy implemented through different parameter space constructions and Bayesian searches [36,41]. This approach is based on the following three key aspects:

First, an appropriate search space was constructed based on the characteristics of the parameters. Linear or logarithmic scales were used for continuous parameters, depending on the magnitude range. For example, parameters that spanned multiple orders of magnitude, such as the learning rate, were represented using log-uniform sampling. In contrast, parameters with narrower ranges, such as subsamples, were sampled uniformly.

Second, the method accounts for logical dependencies between the parameters. In the event of logically incompatible parameter combinations, invalid combinations can be avoided by introducing conditional branches into the search space. Additionally, tree-structured Parzen estimator (TPE) methods [21,42] have been used to efficiently handle these conditional hyperparameter spaces, particularly for tree architecture optimization.

Finally, Bayesian optimization [43] was used to guide the search process. The Bayesian approach was distinguished from traditional grid or random search by constructing probabilistic models between parameters and model performance and by using historical search results to guide subsequent exploration. The pseudocode for the algorithm design is shown in TEXT S7.

2.4. *Model Interpretability Using SHAP Analysis*

The interpretation framework quantified the contribution of each feature to the prediction using the Shapley value [44,45] as a core indicator. A three-level calculation strategy was implemented for all tree models. First, we attempted to invoke the model's native methods to obtain the marginal contribution value of the features in each sample directly. If unavailable, we applied TreeExplainer [46], which utilizes the split paths inside the tree structure and leaf weights to obtain an accurate Shapley value through combined computation. This greatly improved the efficiency compared to the violent computations. When the first two methods proved ineffective, the system transitioned to kernelExplainer, a kernel-function-based approximation computation method [47], which approximates the behavior of a complex model by constructing a local linear model and optimizes the distribution of background samples through k-means clustering.

2.5. *Establishment of Empirical Simulation Equations for Predicting log RCF*

We used three different symbolic regression approaches to develop interpretable prediction equations that could provide insights into the mechanisms underlying PFAS uptake by plants. These methods automatically discover mathematical relationships from data while maintaining physical and chemical interpretability, which is crucial for understanding how molecular properties influence bioaccumulation processes.

2.5.1. Genetic Programming (GP) Symbolic Regression

This model applies GP principles [48] to represent mathematical expressions as tree structures and automatically constructs optimal equations through simulated natural selection. We built expression trees using a function set and optimized candidate solutions through evolutionary operations (crossover, mutation, and elevation). The implementation used the gplearn library with a population size of 2000 and 100 generations and parallel processing to efficiently explore the solution space.

2.5.2. Multilayer Feature Transfer Equation Construction (MFTEC)

MFTEC prioritizes chemically meaningful features before constructing the derived interactions. First, the molecular descriptors that are most relevant to RCFs were identified. We then explored how these key properties interacted to influence plant uptake. The derived interactions in MFTEC have a clearer physical meaning because they are built from preselected chemically relevant features. This approach ensures that the derived features correspond to the actual physicochemical phenomena rather than to mathematical artifacts.

2.5.3. High-Dimensional Sparse Interaction Equation (HSIE)

The HSIE comprehensively explores the molecular property interactions that may influence the RCF before selecting the most important ones. This method can capture complex multifactor effects that are common in environmental systems. By considering higher-order interactions, the HSIE can identify when combinations of molecular properties create uptake behaviors that differ from simple additive effects.

The three approaches differ in how they handle the derived interactions and their physical interpretability. GP can discover any mathematical relationship but may create derived terms that lack environmental relevance. MFTEC builds interactions from chemically important features, ensuring that derived terms like (hydrophobicity $\times$ surface area) represent meaningful physicochemical processes. The HSIE initially considers all possible interactions and then identifies which combinations truly matter for RCF prediction.

The interactions identified by these methods provide insights into the mechanistic basis of plant PFAS uptake. Simple additive models assume that molecular properties independently influence RCF; however, real environmental systems often exhibit synergistic or antagonistic effects. These nonadditive effects appear as interaction terms in the equations. These symbolic regression approaches provide complementary strategies for understanding how the molecular structure governs PFAS root concentration factors. This mechanistic understanding is essential for environmental risk assessment, as it allows for the prediction of the uptake potential of new PFAS compounds based solely on their molecular structure without requiring extensive experimental testing. A flowchart of the study is shown in Figure 1.

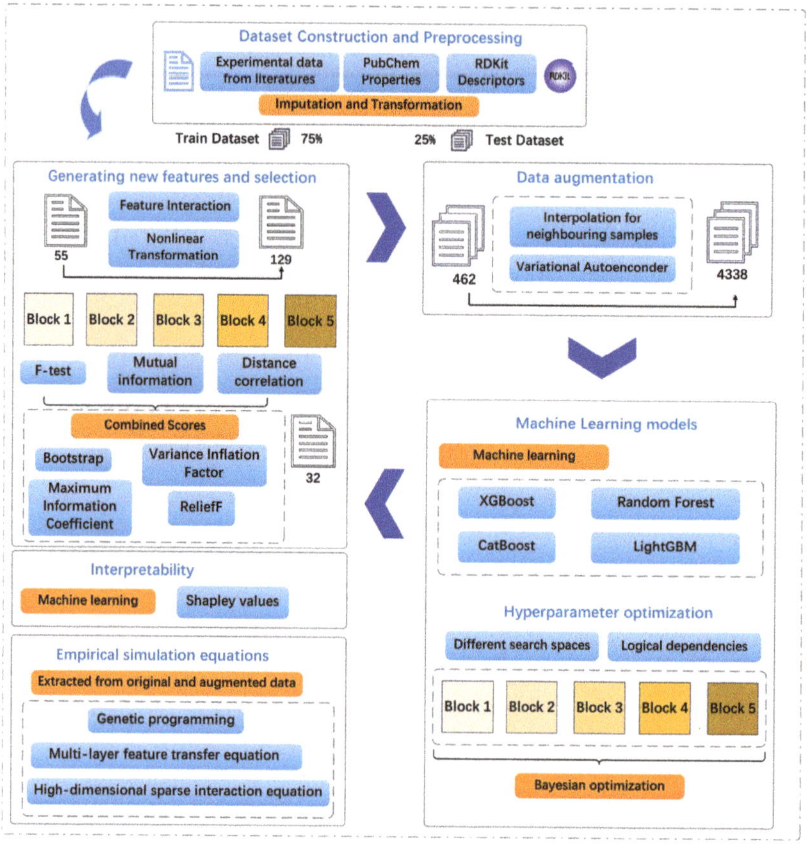

Figure 1. Schematic diagram of machine learning (ML) applied in this work.

3. Results and Discussion

3.1. Based and Derived Features of PFASs Constructed Based on Empirical Formula Performance and Selection

Striking relationships were identified between the key structural and physicochemical features of the PFAS by examining the correlation matrix. The 32 features obtained through feature engineering are shown in Figure 2. This demonstrated that the molecular weight of PFASs emerged as the most important feature in predicting translocation behavior in plants, exhibiting a correlation of 0.69 with the log RCF and a composite score of 0.98. logKow exhibited a moderate positive correlation (0.57) with log RCF, with a characteristic significance of 0.02 and a composite score of 0.74, indicating that the hydrophobicity of PFASs significantly affected their transport behavior. Conversely, the water solubility index, which is based on the topological polar surface area (TPSA)/(Molecular Weight $\times$ (1 + 0.5 $\times$ log Kow)), is a composite feature. This index had a significant predictive value (feature significance of 0.26) and integrated multiple key parameters that affect the water solubility of PFASs. The data substantiated this feature's robust negative correlation (−0.58) with the target variable, indicating that higher-molecular-weight PFASs with lower water solubility accumulate more easily in plant roots [49,50]. In addition, particular consideration should be given to the interaction of exposure time when designing kinetic features. Absorption kinetics was based on the following equation: (1 − exp (−0.005 $\times$ Exposure time)). This is further supported by the correlation of −0.42

with the target variable and a composite score of 0.56. This suggests that this feature captured the time-dependent and hydrophobic barriers to PFAS absorption. In summary, the advantages of molecular weight as a predictor, combined with the significant contributions of hydrophobicity and water solubility indices, indicate that PFAS transport in hydroponic plant systems follows a visible structure–activity relationship. By elucidating these structure-dependent transport mechanisms, we can better predict PFAS's environmental mobility and bioaccumulation potential.

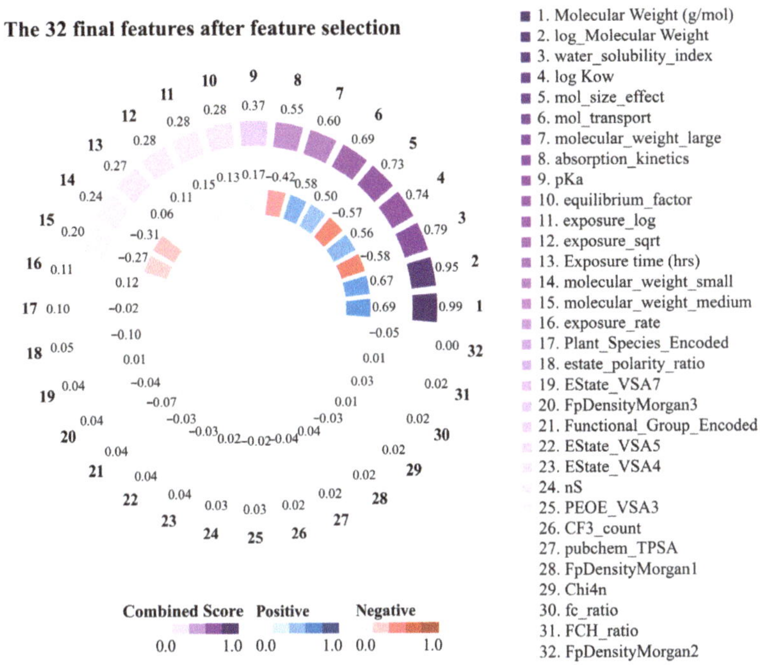

Figure 2. Importance score and correlation with log root concentration factor (RCF) for the top 20 features for root uptake and accumulation of per- and polyfluoroalkyl substances (PFASs) in plants.

3.2. Statistical Analysis Between Original and Augmented Data for the Training Set

The augmentation method used in this study addresses the core challenges of regression modeling scenarios with limited sample sizes and complex data distributions. A statistical comparison of the target variables before and after augmentation is presented in Table S4. The structure of the VAE model used in the enhancement method is shown in Figure 3a. This model achieved effective dimensionality reduction and retained key information regarding the data features. It also enhanced the generation ability and data diversity of the model using a random sampling mechanism of mean and variance. The t-distributed stochastic neighbor embedding (t-SNE) [51] feature space projection diagram (Figure 3b) revealed the topological preservation ability and local density optimization effect of the data augmentation algorithm. The multiple discrete cluster structures formed by the original data (purple) were effectively retained by the enhanced data (orange). Concurrently, the enhanced samples effectively occupied the low-density regions within and surrounding the clusters, thereby markedly enhancing sample coverage without inducing unnatural clusters. The multi-peak nature of the log RCF distribution plot indicated that the target variable may have originated from multiple underlying mechanisms. Furthermore, the enhancement algorithm effectively preserved this intricate distribution structure through adaptive binning [52]. The density curve (Figure 3c) demonstrates that the augmentation process achieved moderate distribution smoothing. The original data (blue) appears to have a more concentrated distribution with a notable peak around −3, while the augmented data (orange) shows a wider, more spread-out distribution with multiple peaks around −5, −3, and closer to −1. This demonstrated that the augmentation process effectively expanded the distribution to cover the entire range of log RCF values, particularly by enhancing the representation of previously underrepresented regions. The histogram (Figure 3d) further illustrates this transformation with the frequency counts before and after augmentation. The orange bars (augmented data) show significantly higher counts across the entire range of log RCF values, with particularly strong representation at the extremes (−6 and −1) and the middle range (−4). This confirmed that the data augmentation strategy successfully increased the sample size while maintaining the general shape of the original distribution but enhanced coverage in previously sparse areas.

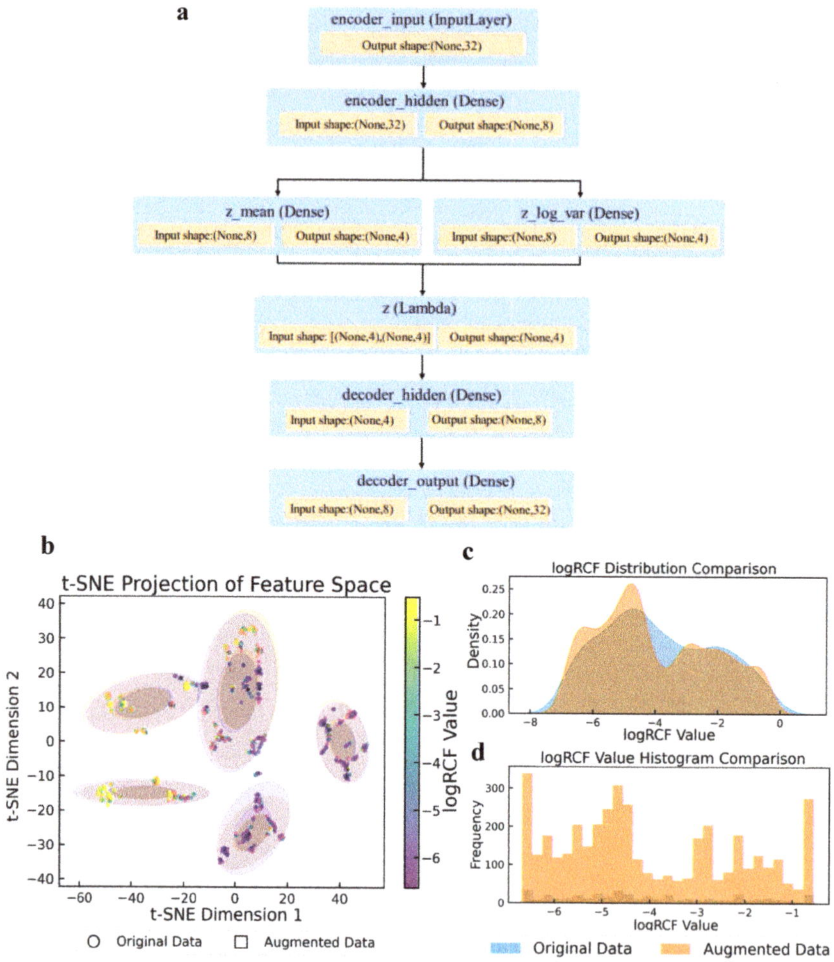

Figure 3. Variational autoencoder (VAE) model structure and comparison of log RCF for PFASs in plants before and after data augmentation. (**a**) VAE model structure. (**b**) t-Distributed stochastic neighbor embedding (t-SNE) projection of feature spaces. (**c**) Log RCF distribution. (**d**) Log RCF histogram comparison.

3.3. Different Model Predictions for log RCF of PFASs in Plants

The performances of multiple ML models developed through multisource data fusion were evaluated by combining experimental data from the literature, physicochemical properties calculated using the RDkit, and chemical descriptors retrieved from PubChem. The data analysis results (Figure 4 and Table 1) present a performance comparison of the four gradient boosting and ensemble learning models (CatBoost, LightGBM, XGBoost, and RF) for the original and enhanced datasets. Specifically, the RMSE of the CatBoost model decreased from 0.7401 to 0.6906 (an improvement of 6.7%) in the five-fold cross-validation and from 0.8015 to 0.7401 (an improvement of 7.7%) in the test set. Its performance indicators increased from 0.8224 to 0.8564 (cross-validation) and from 0.8012 to 0.8300 (test set), after which the RMSE of the LightGBM model decreased by approximately 7.9% (cross-validation) and 9.9% (test set) and its performance improved by approximately 5.2% (cross-validation) and 5.1% (test set). XGBoost performed better after the data enhancement. The RMSE decreased by 10.8% in the cross-validation and 7.8% in the test set, with performance improvements of approximately 6.9% (in the cross-validation) and 4.1% (in the cross-validation and test sets). Although the RF also improved, the improvement was relatively small. The RMSE of the test set decreased by only 1.9% and the RMSE of the cross-validation decreased by 8.8%. Overall, all models achieved a double improvement after using the enhanced data: reduced prediction errors and enhanced model prediction accuracy, with the CatBoost, LightGBM, and XGBoost models exhibiting superior performance in comparison to the RF model on both the original and augmented sets; the CatBoost model provided the optimal fitting results with R^2 = 0.8300, followed by the LightGBM model with R^2 = 0.8249. This discrepancy can be attributed to the fundamental differences in the model mechanisms. CatBoost, LightGBM, and XGBoost use gradient boosting methods to optimize each iteration's objective function by gradually correcting the previous prediction residuals [53]. This allowed them to capture the complex interactions and nonlinear relationships between features more precisely. This characteristic affords gradient boosting methods a notable advantage when confronted with subtle data patterns, culminating in substantial enhancements in the evaluation metrics. Conversely, RF predominantly relies on the independent construction of multiple decision trees and subsequent result averaging to reduce model variance and lacks a continuous correction mechanism for incorrect predictions [28]. A detailed comparison of the predictive results of the four models is shown in Figure S2.

Table 1. Key parameter results of diverse models for root uptake and accumulation of PFASs in hydroponics.

Models	Dataset		R^2	RMSE
CatBoost	Validation	Original	0.8224	0.7401
		Augmented	0.8564	0.6906
	Test	Original	0.8012	0.8015
		Augmented	0.8300	0.7401
LightGBM	Validation	Original	0.8032	0.7793
		Augmented	0.8449	0.7178
	Test	Original	0.7851	0.8334
		Augmented	0.8249	0.7512

Table 1. *Cont.*

Models	Dataset		R^2	RMSE
XGBoost	Validation	Original	0.7953	0.7902
		Augmented	0.8503	0.7050
	Test	Original	0.7827	0.8380
		Augmented	0.8147	0.7727
RandomForest	Validation	Original	0.7913	0.8029
		Augmented	0.8386	0.7321
	Test	Original	0.7713	0.8597
		Augmented	0.7790	0.8438

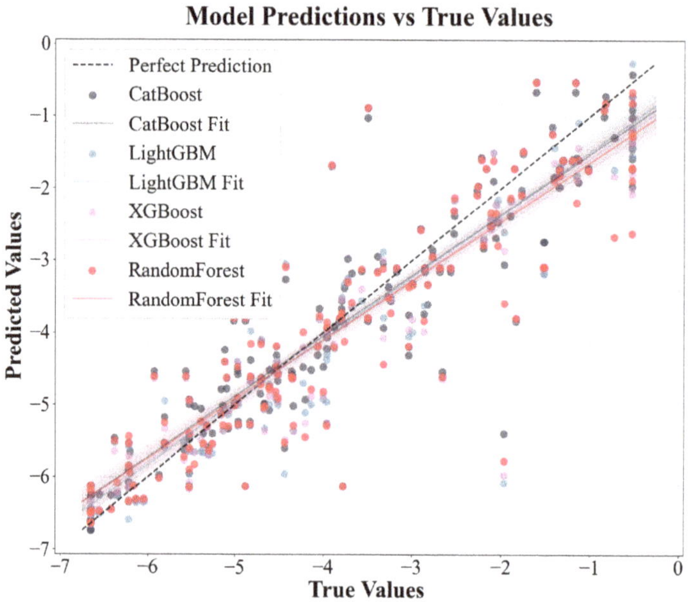

Figure 4. Predicted data vs. original experimental data for four ML models for log RCF.

3.4. *Identification of Different Important Features for Different Predictive Models of log RCF*

Figure 5a–d shows that the molecular structure and physicochemical properties were the dominant predictors across all models. Specifically, plant species and molecular weight-related features demonstrated consistently high absolute SHAP values, indicating their strong influence on the predictions. Furthermore, the exposure time and pKa ranked relatively high as secondarily important factors in all the models. In contrast, the degree of contribution of the electrotopological state (estate)-related characteristics was the lowest in all models. In addition, the different models leverage features in distinct ways. CatBoost relies on unique ordered target statistics and a symmetrical tree structure. In terms of molecular weight characteristics, the value range was moderate (−0.7 to 0.8), and the point distribution was dense and orderly. The feature SHAP distribution range of LightGBM was the widest, especially showing a significant color gradient change in the water solubility index (−1.1 to 1.2), reflecting its tendency to maximize the information gain of a single feature based on the leaf-first strategy. XGBoost exhibited a characteristic distribution between the two. The random model with the lowest prediction accuracy exhibited the most extreme feature differentiation. The SHAP values of the plant species features were abnormally dispersed (−1.0 to 1.3), and the low-ranked features had almost no SHAP contribution. This "black-or-white" evaluation model stems from the simple averaging mechanism of its basic decision tree and the random feature subsampling strategy. The lack of fine adjustment ability of the gradient optimization process for the relationships between subtle features resulted in fewer gradient transitions and a greater aggregation of extreme values, as shown in Figure 5d. An analysis of the feature-importance results for the four models is shown in Figure 5e. All four ML models in this study were assigned extremely high importance scores for molecular weight-related features, which were consistent with the results of the SHAP analysis. In addition, the reliance of the RF on the water solubility index remained the most extreme (0.73), much higher than its assessment of molecular weight (0.32). This pattern, which is overly dominated by a single feature, may cause it to ignore the multifactor synergy effect, thereby becoming a possible reason for its low prediction accuracy. These findings revealed unique feature interpretation perspectives using different algorithmic architectures. This provided a multidimensional empirical basis for understanding the complex relationships between molecular properties and biological effects in prediction models.

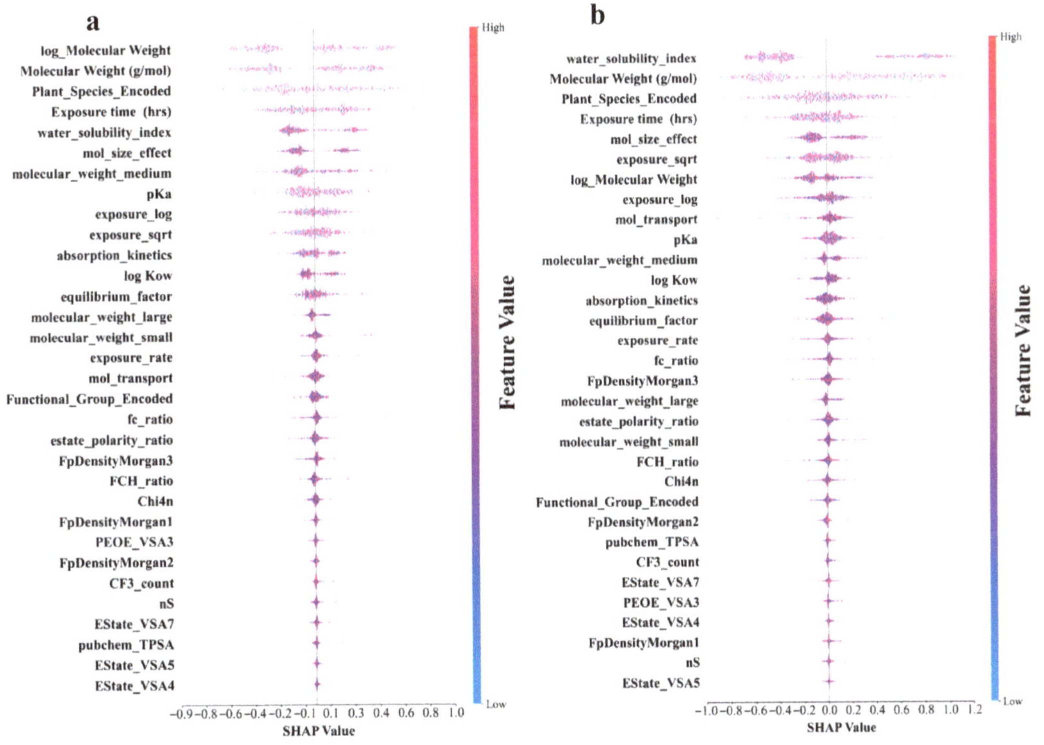

Figure 5. *Cont.*

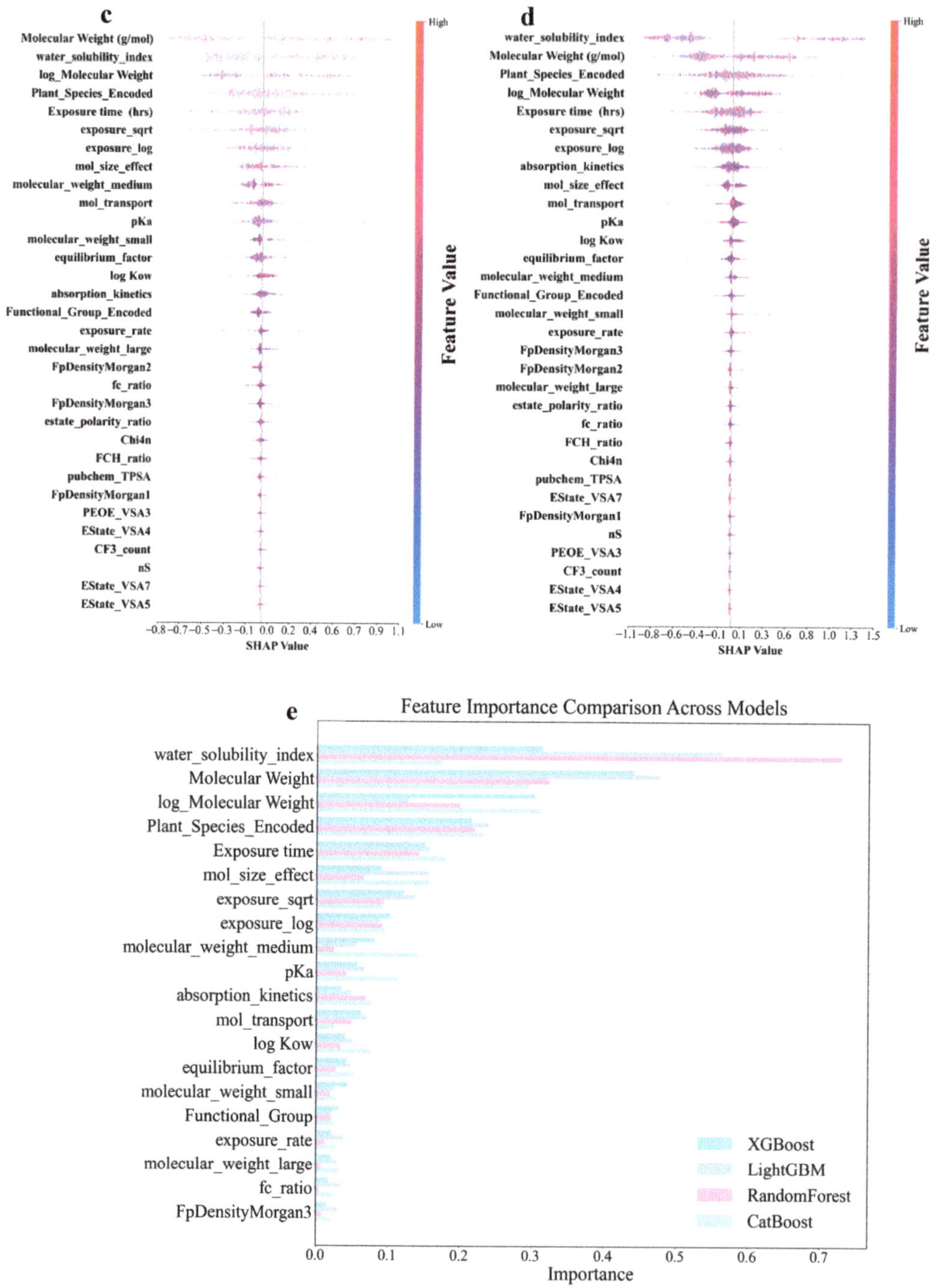

Figure 5. Shapley Additive Explanation (SHAP) values and feature-importance comparison of original and derived features in four ML models for log RCF of PFASs in hydroponics. (**a**) Categorical boosting (CatBoost). (**b**) Light gradient-boosting machine (LightGBM). (**c**) Gradient boosting (XGBoost). (**d**) Random forest. (**e**) Top 20 features comparison for predicting log RCF using four ML models.

3.5. Developing Mathematical Models to Estimate log RCF Values

We adopted three symbolic regression methods and explored the changes in the prediction accuracy of the data before and after enhancement to quantitatively analyze the factors influencing log RCF. The GP model (R^2 = 0.466) for the original data generated highly nonlinear mathematical expressions characterized by complex hierarchical structures and nested functional transformations (Equation (3)). Furthermore, this model (R^2 = 0.556) was used to construct a sophisticated mathematical expression (Equation (4)). This equation incorporates double-nested absolute-value operations around the sine functions. The expression features higher-order exponents of log Kow and pKa, such as $(\log \text{Kow})^6$ and $(\text{pKa})^8$, which mathematically amplify minor variations in these parameters, potentially capturing biological systems in which slight molecular modifications trigger disproportionate bioaccumulation responses. However, the equation's complexity (combining trigonometric, power, square root, and absolute-value operations) creates a "black-box" effect where individual parameter contributions cannot be isolated, limiting mechanistic interpretability. These factors are intertwined through logarithmic operations, power relationships, and multilayer sine functions, thereby reflecting the complexity of the interaction between multiple factors in the uptake of PFASs by plant roots. In summary, incorporating nonlinear and periodic modulation terms in the model underscored the sensitivity of PFAS transport efficiency in plant roots to even minor parameter alterations, potentially corresponding to critical behaviors, such as membrane permeability, ion dissociation, and the balance of intracellular and extracellular distribution within physicochemical processes [50,51,54].

$$\text{logRCF} = \log(0.05) - \sqrt{(\text{logK}_{\text{OW}})^9 + (\text{pKa})^8} + \text{MW} \quad (3)$$

$$\text{logRCF} = \text{MW} + \sin\left(\sin\left(\text{MW} + \left(-0.519 + (\text{logK}_{\text{ow}})^6\right)\right)\right) + \log(0.05) - \sqrt{\sin(\text{MW}) - (\text{pKa})^8} \quad (4)$$

In Equations (3) and (4), 'MW' stands for 'Molecular Weight (g/mol)'.

The MFTEC (R^2 = 0.611) model (Figure 6a) balanced the mathematical complexity with interpretability, producing multivariate polynomial equations with selectively applied transformations (Formula (S1)). The model formed a mathematical structure that combined linear, interactive, and simple nonlinear components. The strongest positive coefficients appeared in exposure metrics (Equilibrium factor: 0.37) and sqrt(|MW|) (0.26), while the most significant adverse effects emerged from pKa (−0.4) and exposure time (−0.34). The augmented data model (R^2 = 0.731) (Figure 6b) was expanded to 46 terms with a sophisticated mathematical structure including quadratic terms, logarithmic transformations, and exponential decay functions (Formula (S2)). The model captures the opposing directional effects of raw and transformed parameters. The inclusion of paired interaction terms with their squared counterparts (MW × sqrt(Exposure time): 1.47 versus ln(MW) × sqrt(Exposure time): −1.02) created mathematical inflection points where the adverse nonlinear effect balanced the positive linear effect.

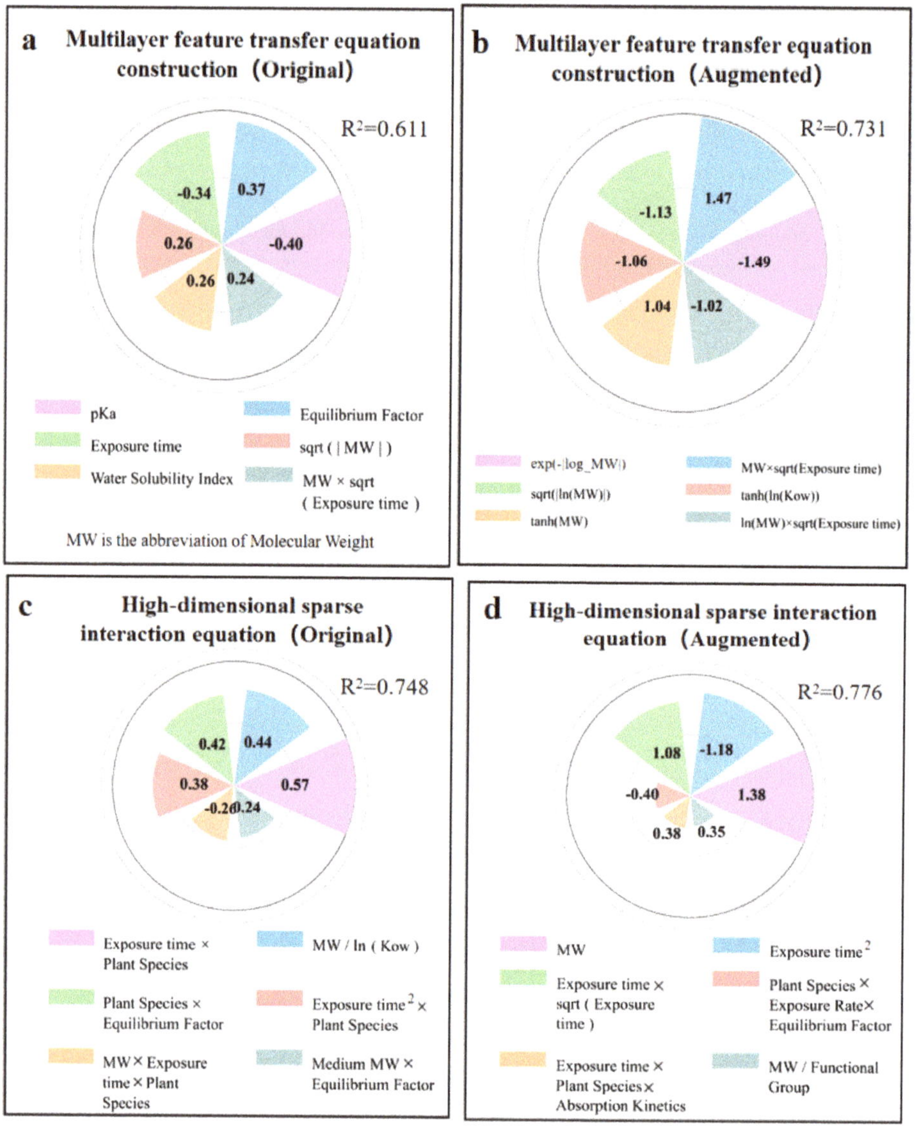

Figure 6. Coefficients from multilayer feature transfer equation construction (MFTEC) and high-dimensional sparse interaction equation (HSIE) in the original and augmented dataset for predicting log RCF. (**a**) Top 6 coefficients for MFTEC in the original data. (**b**) Top 6 coefficients for MFTEC in the augmented data. (**c**) Top 6 coefficients for HSIE in the original data. (**d**) Top 6 coefficients for HSIE in the augmented data.

The HSIE model based on the original data (R^2 = 0.748) (Figure 6c) revealed that the interaction between the exposure time and plant species squared (0.57) had the strongest positive correlation, indicating that plant characteristics exhibited nonlinear cumulative effects during the exposure periods. The molecular weight-to-log Kow ratio (0.44) was the second most influential factor, demonstrating that bioaccumulation followed scale-invariant relationships based on relative molecular properties rather than absolute values. The molecular-weight interaction between exposure time and plant species (−0.26) exhibited a negative correlation, indicating that larger molecules showed reduced bioaccumulation efficiency in certain plant-exposure combinations. The HSIE model, based on enhanced data (R^2 = 0.776) (Figure 6d), showed improved predictive performance through refined parameterization. The molecular-weight term (1.38) became the dominant positive factor, indicating that absolute molecular size plays a crucial role in bioaccumulation. The opposing trends between the negative exposure time-squared term (−1.18) and positive exposure time-square root interaction (1.08) revealed a biphasic bioaccumulation pattern in which the initial rapid uptake transitions to diminishing root concentrations due to translocation to aerial plant parts during prolonged exposure. The interaction between exposure time, plant species, and absorption kinetics (0.38) demonstrated that specific plant species developed enhanced absorption efficiency over extended exposure periods, likely reflecting adaptive physiological responses such as increased membrane permeability or upregulated transporter activity.

Overall, GP had the poorest performance owing to its tree-structure expression generation method based on evolutionary algorithms, which led to overly complex formulas and formed a black-box effect. MFTEC strikes a balance between complexity and interpretability. It first selects important features and then creates derivative features. Moreover, it adopts a stacked integration architecture combined with multiple regressions to form a polynomial equation structure, enabling it to deliver better performance at medium complexity. However, its limitation lies in potentially missing some complex nonlinear relationships that do not fit well within its predetermined transformation functions, particularly when dealing with parameters that exhibit threshold effects or non-monotonic responses. The HSIE adopts the opposite strategy. It first creates derivative features and then selects. This ensures that all possible mathematical relationships are captured, enabling the HSIE to achieve optimal predictive performance by efficiently processing high-dimensional data that contain third-order polynomial features and transformation functions. Its main limitations are the computational intensity required for feature generation and the risk of overfitting when applied to smaller datasets, which can capture noise rather than true biological relationships when there are insufficient data points. The differences between these three methods indicated that balancing the sequence of feature generation and selection, handling nonlinear relationships, and maintaining the interpretability of the equation are the key factors affecting the performance of the final model when constructing a prediction model.

These results provide quantitative support for elucidating the enrichment mechanisms of PFASs in hydroponically grown plants. They emphasized the need to consider the complex interactive effects among the influencing factors when developing environmental risk assessments and pollution remediation strategies.

4. Conclusions

This study established a set of methods based on data-enhanced ML and symbolic regression to evaluate the key factors and quantitative relationships between the bioaccumulation of PFASs in plants. The predictive equations obtained from the three symbolic regression methods provide a unique opportunity to estimate potential PFAS accumulation in plant roots. In conclusion, the findings of this study demonstrate the potential of data augmentation, ML, and symbolic regression as valuable tools for predicting the uptake and accumulation of PFASs in plants. First, we used an existing PFAS dataset related to log RCF for hydroponics. The dataset was augmented by incorporating various methodologies combining SMOTE and VAE under strict quality control to generate a new dataset with 4338 points of experimental data. Additionally, we used four ML techniques on the original and augmented datasets: CatBoost, LightGBM, XGBoost, and RF. Plant species and molecular weight-related features demonstrated consistently high absolute SHAP values, indicating their strong influence on the predictions. Furthermore, the exposure time and pKa ranked relatively high as secondarily important factors in all the models. Finally, through symbolic regression, it was found that the HSIE (R^2 = 0.776) had the best prediction performance. For each model, the prediction accuracy based on the amplified data was higher than that based on the original data. ML models with hyperparameters may be overly complex and difficult to apply in practice. However, mathematical expressions obtained through symbolic regression can compensate for this shortcoming. This is important for sustainable applications in precision agriculture and environmental monitoring. Practical applications of our modeling framework include PFAS risk assessment through rapid RCF prediction, strategic plant selection to minimize root-to-shoot translocation, and agricultural decision support for crop management in contaminated areas. Future research should focus on integrating multi-omics data (transcriptomics, proteomics, and metabolomics) to validate the molecular mechanisms underlying our predictive models and to enhance biological interpretability. In addition, expanding the framework to include plant metabolic transformation pathways and developing real-time monitoring systems for PFAS bioaccumulation in agricultural settings would further advance the practical applications of this methodology. Additionally, it is imperative to emphasize that while ML is poised to serve as a powerful tool for environmental investigation, implementing ML is not a standalone solution, but rather a complement to extensive model interpretation. This approach is essential to elucidate the underlying mechanisms of complex processes, thereby facilitating a more profound understanding of the phenomena under investigation.

Supplementary Materials: The following supporting information can be downloaded at: https://www.mdpi.com/article/10.3390/toxics13070579/s1. TEXT S1: Random forest regressor to impute data. TEXT S2: 1.5 IQR rule. TEXT S3: Assumptions for Generating New Features. TEXT S4: Features correlation. TEXT S5: Features selection. TEXT S6: Data augmentation. TEXT S7: Hyperparameter optimization. Figure S1: Distribution of PFAS Compounds and Plant Species in source dataset. Figure S2: Correlation analysis. Table S1. Statistical summary of the numeric basic variables comprising the PFAS uptake and translocation compiled 616 point dataset. Table S2. Encoded variables. Table S3. Key differences between the four models. Table S4. Statistical comparison of target variable before and after augmentation. Appendix S1. Main python code. References [55,56] are cited in the Supplementary Materials.

Author Contributions: Conceptualization, Y.Z.; Methodology, Y.Z.; Data Curation, Y.Z., Y.L. (Yanting Li), and Y.L. (Yang Li); Writing—Original Draft, Y.Z.; Supervision, L.Z. and Y.Y.; Writing—Review and Editing, L.Z. and Y.Y.; Validation, Y.Y.; Investigation, Y.L. (Yanting Li) and Y.L. (Yang Li). All authors have read and agreed to the published version of the manuscript.

Funding: This research was supported by the National Key Research and Development Program of China (grant number 2024YFC3908500).

Institutional Review Board Statement: Not applicable.

Informed Consent Statement: Not applicable.

Data Availability Statement: The data supporting the findings of this study are available from the corresponding author upon reasonable request.

Conflicts of Interest: The authors declare that they have no competing financial interests or personal relationships that may have influenced the work reported in this study.

References

1. Li, X.; Shen, X.; Jiang, W.; Xi, Y.; Li, S. Comprehensive review of emerging contaminants: Detection technologies, environmental impact, and management strategies. *Ecotoxicol. Environ. Saf.* **2024**, *278*, 116420. [CrossRef]
2. Ilango, A.K.; Zhang, W.; Liang, Y. Uptake of per- and polyfluoroalkyl substances by Conservation Reserve Program's seed mix in biosolids-amended soil. *Environ. Pollut.* **2024**, *363*, 125235. [CrossRef] [PubMed]
3. Li, Z. Plant Uptake Models of Pesticides: Advancing Integrated Pest Management, Food Safety, and Health Risk Assessment. *Rev. Environ. Contam. Toxicol.* **2025**, *263*, 3. [CrossRef]
4. Trapp, S. Modelling uptake into roots and subsequent translocation of neutral and ionisable organic compounds. *Pest Manag. Sci.* **2000**, *56*, 767–778. [CrossRef]
5. Li, Y.; Sallach, J.B.; Zhang, W.; Boyd, S.A.; Li, H. Characterization of Plant Accumulation of Pharmaceuticals from Soils with Their Concentration in Soil Pore Water. *Environ. Sci. Technol.* **2022**, *56*, 9346–9355. [CrossRef] [PubMed]
6. Schymanski, E.L.; Zhang, J.; Thiessen, P.A.; Chirsir, P.; Kondic, T.; Bolton, E.E. Per- and Polyfluoroalkyl Substances (PFAS) in PubChem: 7 Million and Growing. *Environ. Sci. Technol.* **2023**, *57*, 16918–16928. [CrossRef]
7. Evich, M.G.; Davis, M.J.B.; McCord, J.P.; Acrey, B.; Awkerman, J.A.; Knappe, D.R.U.; Lindstrom, A.B.; Speth, T.F.; Tebes-Stevens, C.; Strynar, M.J.; et al. Per- and polyfluoroalkyl substances in the environment. *Science* **2022**, *375*, eabg9065. [CrossRef]
8. Ogunbiyi, O.D.; Ajiboye, T.O.; Omotola, E.O.; Oladoye, P.O.; Olanrewaju, C.A.; Quinete, N. Analytical approaches for screening of per- and poly fluoroalkyl substances in food items: A review of recent advances and improvements*. *Environ. Pollut.* **2023**, *329*, 121705. [CrossRef]
9. Ji, Y.; Wang, X.; Wang, R.; Wang, J.; Zhao, X.; Wu, F. Toxicity prediction and risk assessment of per- and polyfluoroalkyl substances for threatened and endangered fishes. *Environ. Pollut.* **2024**, *361*, 124920. [CrossRef]
10. Schriever, C.; Lamshoeft, M. Lipophilicity matters—A new look at experimental plant uptake data from literature. *Sci. Total Environ.* **2020**, *713*, 136667. [CrossRef]
11. Yang, H.; Zhao, Y.; Chai, L.; Ma, F.; Yu, J.; Xiao, K.-Q.; Gu, Q. Bio-accumulation and health risk assessments of per- and polyfluoroalkyl substances in wheat grains. *Environ. Pollut.* **2024**, *356*, 124351. [CrossRef] [PubMed]
12. Ismail, U.M.; Elnakar, H.; Khan, M.F. Sources, Fate, and Detection of Dust-Associated Perfluoroalkyl and Polyfluoroalkyl Substances (PFAS): A Review. *Toxics* **2023**, *11*, 335. [CrossRef] [PubMed]
13. Nayak, S.; Sahoo, G.; Das, I.I.; Mohanty, A.K.; Kumar, R.; Sahoo, L.; Sundaray, J.K. Poly- and Perfluoroalkyl Substances (PFAS): Do They Matter to Aquatic Ecosystems? *Toxics* **2023**, *11*, 543. [CrossRef]
14. Collins, C.D.; Finnegan, E. Modeling the Plant Uptake of Organic Chemicals, Including the Soil-Air-Plant Pathway. *Environ. Sci. Technol.* **2010**, *44*, 998–1003. [CrossRef] [PubMed]
15. Qu, R.; Wang, J.X.; Li, X.J.; Zhang, Y.; Yin, T.L.; Yang, P. Per- and Polyfluoroalkyl Substances (PFAS) Affect Female Reproductive Health: Epidemiological Evidence and Underlying Mechanisms. *Toxics* **2024**, *12*, 678. [CrossRef]
16. Alnaimat, S.; Mohsen, O.; Elnakar, H. Perfluorooctanoic Acids (PFOA) removal using electrochemical oxidation: A machine learning approach. *J. Environ. Manag.* **2024**, *370*, 122857. [CrossRef]
17. Nie, Q.; Liu, T. Large language models: Tools for new environmental decision-making. *J. Environ. Manag.* **2025**, *375*, 124373. [CrossRef]
18. Gao, F.; Shen, Y.; Sallach, B.; Li, H.; Zhang, W.; Li, Y.; Liu, C. Predicting crop root concentration factors of organic contaminants with machine learning models. *J. Hazard. Mater.* **2022**, *424*, 127437. [CrossRef]
19. Wang, Y.; Dong, J.; Zhou, Y.; Cheng, Y.; Zhao, X.; Peijnenburg, W.J.G.M.; Vijver, M.G.; Leung, K.M.Y.; Fan, W.; Wu, F. Addressing the Data Scarcity Problem in Ecotoxicology via Small Data Machine Learning Methods. *Environ. Sci. Technol.* **2025**, *59*, 5867–5871. [CrossRef]
20. Rao, K.M.; Saikrishna, G.; Supriya, K. Data preprocessing techniques: Emergence and selection towards machine learning models-a practical review using HPA dataset. *Multimed. Tools Appl.* **2023**, *82*, 37177–37196. [CrossRef]
21. Reddy, G. A reinforcement-based mechanism for discontinuous learning. *Proc. Natl. Acad. Sci. USA* **2022**, *119*, e2215352119. [CrossRef] [PubMed]
22. Moritz, P.; Nishihara, R.; Wang, S.; Tumanov, A.; Liaw, R.; Liang, E.; Elibol, M.; Yang, Z.; Paul, W.; Jordan, M.I.; et al. Ray: A Distributed Framework for Emerging AI Applications. *arXiv* **2018**, arXiv:1712.05889.
23. Liu, S.; Kappes, B.B.; Amin-ahmadi, B.; Benafan, O.; Zhang, X.; Stebner, A.P. Physics-informed machine learning for composition–process–property design: Shape memory alloy demonstration. *Appl. Mater. Today* **2021**, *22*, 100898. [CrossRef]
24. Maeda, K.; Hirano, M.; Hayashi, T.; Iida, M.; Kurata, H.; Ishibashi, H. Elucidating Key Characteristics of PFAS Binding to Human Peroxisome Proliferator-Activated Receptor Alpha: An Explainable Machine Learning Approach. *Environ. Sci. Technol.* **2023**, *58*, 488–497. [CrossRef]
25. Ileberi, E.; Sun, Y.; Wang, Z. A machine learning based credit card fraud detection using the GA algorithm for feature selection. *J. Big Data* **2022**, *9*, 24. [CrossRef]
26. Song, W.C.; Xie, J. Group feature screening via the F statistic. *Commun. Stat.-Simul. Comput.* **2022**, *51*, 1921–1931. [CrossRef]

27. Gonzalez, M.E.; Silva, J.F.; Videla, M.; Orchard, M.E. Data-Driven Representations for Testing Independence: Modeling, Analysis and Connection With Mutual Information Estimation. *IEEE Trans. Signal Process.* **2022**, *70*, 158–173. [CrossRef]
28. Edelmann, D.; Mori, T.F.; Szekely, G.J. On relationships between the Pearson and the distance correlation coefficients. *Stat. Probab. Lett.* **2021**, *169*, 108960. [CrossRef]
29. Xu, P.; Nian, M.; Xiang, J.; Zhang, X.; Cheng, P.; Xu, D.; Chen, Y.; Wang, X.; Chen, Z.; Lou, X.; et al. Emerging PFAS Exposure Is More Potent in Altering Childhood Lipid Levels Mediated by Mitochondrial DNA Copy Number. *Environ. Sci. Technol.* **2025**, *59*, 2484–2493. [CrossRef]
30. Wang, Z.; Zhang, J.; Dai, Y.; Zhang, L.; Guo, J.; Xu, S.; Chang, X.; Wu, C.; Zhou, Z. Mediating effect of endocrine hormones on association between per- and polyfluoroalkyl substances exposure and birth size: Findings from sheyang mini birth cohort study. *Environ. Res.* **2023**, *226*, 115658. [CrossRef]
31. Santibanez, N.; Vega, M.; Perez, T.; Enriquez, R.; Escalona, C.E.; Oliver, C.; Romero, A. In vitro effects of phytogenic feed additive on *Piscirickettsia salmonisgrowth* and biofilm formation. *J. Fish Dis.* **2024**, *47*, e13913. [CrossRef] [PubMed]
32. Bolon-Canedo, V.; Sanchez-Marono, N.; Alonso-Betanzos, A. Recent advances and emerging challenges of feature selection in the context of big data. *Knowl.-Based Syst.* **2015**, *86*, 33–45. [CrossRef]
33. Kovacs, K.D.; Haidu, I. Tracing out the effect of transportation infrastructure on NO_2 concentration levels with Kernel Density Estimation by investigating successive COVID-19-induced lockdowns. *Environ. Pollut.* **2022**, *309*, 119719. [CrossRef]
34. Dong, J.; Tsai, G.; Olivares, C.I. Prediction of 35 Target Per- and Polyfluoroalkyl Substances (PFASs) in California Groundwater Using Multilabel Semisupervised Machine Learning. *ACS EST Water* **2023**, *4*, 969–981. [CrossRef]
35. Kang, J.-K.; Lee, D.; Muambo, K.E.; Choi, J.-w.; Oh, J.-E. Development of an embedded molecular structure-based model for prediction of micropollutant treatability in a drinking water treatment plant by machine learning from three years monitoring data. *Water Res.* **2023**, *239*, 120037. [CrossRef] [PubMed]
36. Ng, K.; Alygizakis, N.; Androulakakis, A.; Galani, A.; Aalizadeh, R.; Thomaidis, N.S.; Slobodnik, J. Target and suspect screening of 4777 per- and polyfluoroalkyl substances (PFAS) in river water, wastewater, groundwater and biota samples in the Danube River Basin. *J. Hazard. Mater.* **2022**, *436*, 129276. [CrossRef]
37. Lyu, H.; Xu, Z.; Zhong, J.; Gao, W.; Liu, J.; Duan, M. Machine learning-driven prediction of phosphorus adsorption capacity of biochar: Insights for adsorbent design and process optimization. *J. Environ. Manag.* **2024**, *369*, 122405. [CrossRef]
38. Zheng, S.-S.; Guo, W.-Q.; Lu, H.; Si, Q.-S.; Liu, B.-H.; Wang, H.-Z.; Zhao, Q.; Jia, W.-R.; Yu, T.-P. Machine learning approaches to predict the apparent rate constants for aqueous organic compounds by ferrate. *J. Environ. Manag.* **2023**, *329*, 116904. [CrossRef]
39. Lee, E.; You, Y.-W.; Jung, Y.-H.; Kam, J. Explainable AI-based risk assessment for pluvial floods over South Korea. *J. Environ. Manag.* **2025**, *385*, 125640. [CrossRef]
40. Li, T.; Wu, Y.; Ren, F.; Li, M. Estimation of unrealized forest carbon potential in China using time-varying Boruta-SHAP-random forest model and climate vegetation productivity index. *J. Environ. Manag.* **2025**, *377*, 124649. [CrossRef]
41. Ding, J.; Lee, S.-J.; Vlahos, L.; Yuki, K.; Rada, C.C.; van Unen, V.; Vuppalapaty, M.; Chen, H.; Sura, A.; McCormick, A.K.; et al. Therapeutic blood-brain barrier modulation and stroke treatment by a bioengineered FZD_4-selective WNT surrogate in mice. *Nat. Commun.* **2023**, *14*, 2947. [CrossRef]
42. Cao, H.; Peng, J.; Zhou, Z.; Sun, Y.; Wang, Y.; Liang, Y. Insight into the defluorination ability of per- and polyfluoroalkyl substances based on machine learning and quantum chemical computations. *Sci. Total Environ.* **2022**, *807*, 151018. [CrossRef] [PubMed]
43. Schossler, R.T.; Ojo, S.; Yu, X.B. Optimizing Photodegradation Rate Prediction of Organic Contaminants: Models with Fine-Tuned Hyperparameters and SHAP Feature Analysis for Informed Decision Making. *ACS EST Water* **2023**, *4*, 1131–1145. [CrossRef]
44. Sheik, A.G.; Krishna, S.B.N.; Patnaik, R.; Ambati, S.R.; Bux, F.; Kumari, S. Digitalization of phosphorous removal process in biological wastewater treatment systems: Challenges, and way forward. *Environ. Res.* **2024**, *252*, 119133. [CrossRef] [PubMed]
45. Fabregat-Palau, J.; Ershadi, A.; Finkel, M.; Rigol, A.; Vidal, M.; Grathwohl, P. Modeling PFAS Sorption in Soils Using Machine Learning. *Environ. Sci. Technol.* **2025**, *59*, 7678–7687. [CrossRef] [PubMed]
46. Lundberg, S.M.; Erion, G.; Chen, H.; DeGrave, A.; Prutkin, J.M.; Nair, B.; Katz, R.; Himmelfarb, J.; Bansal, N.; Lee, S.-I. From local explanations to global understanding with explainable AI for trees. *Nat. Mach. Intell.* **2020**, *2*, 56–67. [CrossRef]
47. Xu, P.; Peng, J.; Yuan, T.; Chen, Z.; He, H.; Wu, Z.; Li, T.; Li, X.; Wang, L.; Gao, L.; et al. High-throughput mapping of single-neuron projection and molecular features by retrograde barcoded labeling. *eLife* **2024**, *13*, e85419. [CrossRef]
48. Pak, W.; Hindges, R.; Lim, Y.S.; Pfaff, S.L.; O'Leary, D.D.M. Magnitude of binocular vision controlled by islet-2 repression of a genetic program that specifies laterality of retinal axon pathfinding. *Cell* **2004**, *119*, 567–578. [CrossRef]
49. Adu, O.; Bryant, M.T.; Ma, X.; Sharma, V.K. A Machine Learning Approach for Predicting Plant Uptake and Translocation of Per- and Polyfluoroalkyl Substances (PFAS) from Hydroponics. *ACS EST Eng.* **2024**, *4*, 1884–1890. [CrossRef]
50. Huang, D.; Xiao, R.; Du, L.; Zhang, G.; Yin, L.; Deng, R.; Wang, G. Phytoremediation of poly- and perfluoroalkyl substances: A review on aquatic plants, influencing factors, and phytotoxicity. *J. Hazard. Mater.* **2021**, *418*, 126314. [CrossRef]
51. Linderman, G.C.; Rachh, M.; Hoskins, J.G.; Steinerberger, S.; Kluger, Y. Fast interpolation-based t-SNE for improved visualization of single-cell RNA-seq data. *Nat. Methods* **2019**, *16*, 243–245. [CrossRef] [PubMed]

52. Chen, K.; Zhang, Z. Pedestrian Counting with Back-Propagated Information and Target Drift Remedy. *IEEE Trans. Syst. Man Cybern. Syst.* **2017**, *47*, 639–647. [CrossRef]
53. Nguyen, D.V.; Seo, M.; Chen, Y.; Wu, D. Enhancing hydrogen sulfide control in urban sewer systems using machine learning models: Development of a new predictive simulation approach by using boosting algorithm. *J. Hazard. Mater.* **2025**, *491*, 137906. [CrossRef] [PubMed]
54. Li, F.; Duan, J.; Tian, S.; Ji, H.; Zhu, Y.; Wei, Z.; Zhao, D. Short-chain per- and polyfluoroalkyl substances in aquatic systems: Occurrence, impacts and treatment. *Chem. Eng. J.* **2020**, *380*, 122506. [CrossRef]
55. Montal, M.; Mueller, P. Formation of bimolecular membranes from lipid monolayers and a study of their electrical properties. *Proc. Natl. Acad. Sci. USA* **1972**, *69*, 3561–3566. [CrossRef]
56. Potts, D.S.; Bregante, D.T.; Adams, J.S.; Torres, C.; Flaherty, D.W. Influence of solvent structure and hydrogen bonding on catalysis at solid-liquid interfaces. *Chem. Soc. Rev.* **2021**, *50*, 12308–12337. [CrossRef]

MDPI AG
Grosspeteranlage 5
4052 Basel
Switzerland
Tel.: +41 61 683 77 34

Toxics Editorial Office
E-mail: toxics@mdpi.com
www.mdpi.com/journal/toxics

www.ingramcontent.com/pod-product-compliance
Lightning Source LLC
Chambersburg PA
CBHW061628210926
51850CB00009B/2468